T0219960

Springer Praxis Books

Popular Science

This book series presents the whole spectrum of Earth Sciences, Astronautics and Space Exploration. Practitioners will find exact science and complex engineering solutions explained scientifically correct but easy to understand. Various subseries help to differentiate between the scientific areas of Springer Praxis books and to make selected professional information accessible for you.

The Springer Praxis Popular Science series contains fascinating stories from around the world and across many different disciplines. The titles in this series are written with the educated lay reader in mind, approaching nitty-gritty science in an engaging, yet digestible way. Authored by active scholars, researchers, and industry professionals, the books herein offer far-ranging and unique perspectives, exploring realms as distant as Antarctica or as abstract as consciousness itself, as modern as the Information Age or as old our planet Earth. The books are illustrative in their approach and feature essential mathematics only where necessary. They are a perfect read for those with a curious mind who wish to expand their understanding of the vast world of science.

More information about this series at
http://www.springer.com/series/4097

Sergio Rossi

A Journey in Antarctica

Exploring the Future of the White Continent

Sergio Rossi
University of Salento
Lecce, Italy

Universidade Federal do Ceará
Fortaleza, Brazil

ISSN 2626-6113   ISSN 2626-6121   (electronic)
Springer Praxis Books
ISBN 978-3-030-89494-8   ISBN 978-3-030-89492-4   (eBook)
https://doi.org/10.1007/978-3-030-89492-4

© The Editor(s) (if applicable) and The Author(s), under exclusive licence to Springer Nature Switzerland AG 2022

This work is subject to copyright. All rights are solely and exclusively licensed by the Publisher, whether the whole or part of the material is concerned, specifically the rights of reprinting, reuse of illustrations, recitation, broadcasting, reproduction on microfilms or in any other physical way, and transmission or information storage and retrieval, electronic adaptation, computer software, or by similar or dissimilar methodology now known or hereafter developed.

The use of general descriptive names, registered names, trademarks, service marks, etc. in this publication does not imply, even in the absence of a specific statement, that such names are exempt from the relevant protective laws and regulations and therefore free for general use.

The publisher, the authors and the editors are safe to assume that the advice and information in this book are believed to be true and accurate at the date of publication. Neither the publisher nor the authors or the editors give a warranty, expressed or implied, with respect to the material contained herein or for any errors or omissions that may have been made. The publisher remains neutral with regard to jurisdictional claims in published maps and institutional affiliations.

This Springer imprint is published by the registered company Springer Nature Switzerland AG.
The registered company address is: Gewerbestrasse 11, 6330 Cham, Switzerland

A book that explains why we should care about what happens on the white continent.

To Lidriana, the architect of a renaissance

Foreword

Why are we going to Antarctica? That was the topic of conversation in the cabin of Wolf Arntz, expedition manager of the *Polarstern* on its 2000 cruise. There were about ten of us gathered on the best scientific icebreaker on the planet, and after an intense day, we had come to that topic because this monster can burn 1,400,000 € in diesel during the two-month voyage beyond where anyone goes: the remote confines of Atka Bay, the Larsen area, or Austasen, the icebergs' resting place. "At the beginning the campaigns were designed to make a rigorous study of the natural resources (fishing, mining, etc.) offered by the white continent," said Arntz, "but soon the scientists realized that there was something else there, the opportunity to study a pristine place undisturbed by the hand of man, and fascinating." We each reflected on this point while sipping our Rioja or Priorat.

"There are certainly various 'levels of interest' in coming here," said Tom Brey, Arntz's then right-hand man and second-in-command for science. Both men worked at Alfred Wegener Institute, Germany's polar research center, generously endowed with monetary and human capital for polar research. "The first level is political," Brey continued. "It's important to be here; that's why Germany (which has not openly declared territorial interests like Britain or Argentina) has a research institute, an unrivalled icebreaker ship and a base like Neumayer, an engineering prodigy. But then there started to be other levels of interest, like the fact that this is a unique place where you can study the evolution of species, ocean–atmosphere interaction, climate change or the atmospheric chemistry of ozone like nowhere else."

The conversation became lively, and I was aware that the aim far exceeded a whim among scientists to travel to the Antipodes and spend a few days watching penguins. "Antarctica is an open book that has not yet received any human impact," stressed my then boss, Josep Maria Gili from the Institute of Marine Sciences of Spain's Institute of Marine Sciences (CSIC): "The history of the planet is written in its waters and we have to take advantage of it."

When in my thirties I was told that I was going to voyage aboard the *Polarstern*, none of these ideas crossed my mind. I was simply to have an adventure on which only a few privileged people had the honor to be invited, and I was going to do so on the best polar oceanographic vessel in the world. However, after three expeditions (in 2000, in 2003/04, and in 2011), I felt the need to explain many things to those who have not been and will never go there. I am not an expert on Antarctica (going on three expeditions does not give you that honor; some have worked for more than thirty years on this subject alone), but above all I needed to reply to the skeptics who see the white continent only as a remote place to which privileged people travel to explore the remote corners of the planet.

"Science is everybody's heritage," Gili once told me; "you owe it to those who pay you." Not all scientists are clear on this, but I am one of those who regard the option of a scientific career as something that is offered by society and, as such, its outcomes should be integrated into it as much as possible. We are neither Dirac nor Marie Curie, and Darwin's trajectory can never again be emulated because a solitary, self-absorbed, brilliant scientist isolated from society can no longer be supported. We are obliged to explain what we do, since our work is of little value if we do not transmit it, if we do not external-ize it at a level of understanding appropriate to a supermarket manager, a busy cab driver, or a lawyer who is enthusiastic about nature.

There is something more. In these critical times when things are changing so fast, it is a matter of great urgency to explain concepts in a clear, concise, entertaining, and rigorous way. We need to convey to people that we are at a crossroads, also what is happening now and what will happen next. If the location is as remote as Antarctica, we have to make an even greater effort to get that message across to as many people as possible.

In view of the role of science communication in these past two decades, I needed to convey why Antarctica is so crucial, why we go there, and why every-thing that happens on that immense and seemingly lifeless block of inert ice can either benefit or harm us. And I found no better way to do so than by compiling as much information as possible to transmit in a personal way on topics that I have either mastered or not, along with stories, reflections, and images. This book is an attempt to provoke opinion on something far away, and readers will be sur-prised to know to what extent the white continent fascinates and interests us, and above all how much our future quality of life depends on what happens there.

The first part of the book explains the basics of the continent, how it works and its geological history, how everything depends on seasonal dynamics, and how the ice has been forming and melting cyclically for tens of millions of years. I then explain how climate change is the key to understanding the future of both Antarctica in particular and the planet in general. We seldom speak of Antarctica, yet it acts as the thermostat of the planet. It contains the planet's greatest amount of ice in the form of huge glaciers, reaching more than 3500 meters deep. Both the Arctic (from *arktos*, bear) and the Antarctic (from *ank-arktos*, without bears) are vital places to understand the future and past climatology of our planet.

I then describe the direct impact from pollution, fishing, mining, and tourism and invasive species, and their convulsive synergy so very present in the evolution of this remote place (and the entire planet). Lastly, I give my impressions of life aboard the vast oceanographic vessel that welcomed me, of the visions that I had the privilege to commit to memory, and of the thoughts that went through my head when I found myself in the most isolated place in the world.

I hope that by the end of this book the reader will have a slightly different view of the white continent, a better concept of what happens there and how it can affect us all, even though so far from home. Perhaps, this is the place where globalization is demonstrated in a palpable way—good for some things and disastrous for others (Fig. 1).

Fig. 1 One of the first icebergs seen on the 2000 *Polarstern* expedition

Lecce, Italy   Sergio Rossi

Contents

1   The Continent at the End of the World   1

2   A Brief History of an Arduous Discovery   7

3   Paleoclimate: Looking for the Importance of Antarctica   13

4   Survival Lessons: Mars on Earth   21

5   Perpetually Moving Glaciers: Where Icebergs Emerge   27

6   The Antarctic Ice Floe   37

7   Microbes   43

8   Ice, Algae, Krill   51

9   Life in the Dark: A Diversity Explosion Under the Ice   57

10 Birds at the End of the World   69

11 Large Mammals of the White Continent   79

12 Climate Change: Not So Isolated   91

13 What Happened in Larsen?   101

14   Organisms and Climate Change   109

15   The Ozone Layer and Antarctica   121

16   Pollutants in a Pristine Place   127

17   The Final Resort   133

18   Remote Fishing Grounds   141

19   Ecotourism and Invasive Species   147

20   The Antarctic Treaty   153

21   Antarctic Bases   161

22   Living on an Ocean-Going Icebreaker   167

23   The Last Bastion of an Unspoiled Planet   173

Epilogue   181

Bibliography   183

List of Figures

Fig. 3.1   The German ship Polarstern, the biggest oceanographic ice
            breaker of the world   14
Fig. 5.1   The sea is continuously interacting with the ice edge, eroding and
            fragmenting the shelf   29
Fig. 5.2   A giant pycnogonid, the sea spider living in the Antarctic benthos   35
Fig. 7.1   Rosette ready to be deployed to sample water and physical
            parameters   45
Fig. 9.1   Sponge grounds in the Antarctic benthos (Photo credits, Julian
            Gutt, AWI)   60
Fig. 10.1   Emperor penguin colony, with adults and chicks   70
Fig. 10.2   The birds may travel thousands of kilometers in search of food   74
Fig. 11.1   Young Weddell Sea seal on the Antarctic ice sheet   88
Fig. 12.1   Swimming pool on an iceberg   93
Fig. 13.1   Larsen A, where the air temperature reached more than 10 °C   103
Fig. 13.2   A rare photo of Larsen C, one of the most unreachable zones on
            Earth   107
Fig. 14.1   The icebreaker *Polarstern* entering the ice sheet at Austasen   111
Fig. 16.1   Green iceberg in the middle of the Weddell Sea   129
Fig. 17.1   Logistics at Jubany, the Argentinian–German Antarctic base   135
Fig. 19.1   Bike used to visit the meteorological station in the German base
            of Neumayer, spotted on 2000 expedition   148
Fig. 20.1   TV grab, especially designed to pick up intact sediment/organism
            samples and explore the bottoms   155
Fig. 21.1   The new Neumayer station in 2011   162
Fig. 22.1   Football at Neumayer station (2003): sailors and technicians vs
            scientists; the first team won the match   171
Fig. 23.1   Infinite ice, at the end of the expedition   174
Fig. 23.2   Blue iceberg, observed at the end of the 2000 expedition   178

1

The Continent at the End of the World

A Very Peculiar Place

Due to the heavy storm that continues to block us in the days after Christmas 2003, the captain of the *Polarstern* has banned the crew from going on deck. I look out of the bridge window of this icebreaker, the pride of the Germans, and realize that, even if I were well sheltered, a man like me might not survive 2 days on deck. The winds exceed a hundred kilometers per hour and the sky is white with a blizzard that barely lets you glimpse the ice floes through the snow and ice blowing horizontally, creating graceful shapes in their path. I have the feeling of being in the most isolated place on the planet, a place where humankind is undoubtedly still a guest and without proper technology can never dominate. In fact, it is the only continent in which we were unable to survive, until the arrival of the seal hunters and whalers in the late nineteenth century and especially the early twentieth century, who tried to stay only in summer because winter overcame them.

Why do I miss it? In Antarctica, it is only since the late 1960s and 70s that perennial settlements have been built, housing between 1000 and 4000 people depending on the time of year. No less than 90% of the so-called cryosphere of the planet (the frozen layer on continents) is on just over 280,000 square kilometers of land (just over half the area of Spain), covering an area of more than 14 million square kilometers (in fact only 0.4% of the continent is dry land, according to the latest available data). This is a strange paradox because the planet's fourth continent, after Asia, America and Africa, is practically all ice, which can be more than 4000 m thick in some points, so the part that is of land and rocks is very small.

© The Author(s), under exclusive license to Springer Nature Switzerland AG 2022
S. Rossi, *A Journey in Antarctica*, Springer Praxis Books,
https://doi.org/10.1007/978-3-030-89492-4_1

The ice is in perpetual movement. The glaciers are 'alive', continuously moving billions of tons of ice from the interior to the coastal zone. In such a place, it should not surprise us that the average temperature never exceeds 0 °C even in the 'warmest' places, such as the northernmost tip of the Antarctic Peninsula, and go down as far as the legendary -89.3 °C recorded at the Russian station of Vostok on 21 July 1983.

I keep looking out the window and, with the approval of the second officer, holding a cup overflowing with (undrinkable) coffee I dare to open the door. As I step outside, I feel the icy wind rushing through my body. Lutz's face breaks into a smile at my sudden astonishment: I'm warm, in a way that's beyond me. "We've reached 140 km/h," he announces when he sees me coming back after no more than 120 s. This is no windspeed at all, compared to over 200 km/h reached in the interior of the continent but, whichever way you look at it, it is certainly brutal.

Antarctica is a special place, where only unique animals and plants can withstand the cold and the long seasons of perpetual light and others of dark night. It is a little strange: during the months of December to January (austral summer) the light is continuous and the sun never sets, so why isn't it warmer? The answer lies in the slant of the rays that reach Antarctica, an oblique light that does not heat nearly as much as the light at the equator or even that in temperate zones such as Spain or France, which are more at the zenith.

In addition, the large amount of ice has an effect known as the albedo, namely reflection of 75 to 80% of the light coming from the sun. Antarctica does not retain heat. We cannot generalize. There are warmer and colder areas within the frozen continent. Living on a coastal base on the Peninsula, where the average temperature is -22 °C in August (austral winter), is not the same as living in the interior, where that average drops to -33 °C in the same month. In fact, for every 1000 m of altitude the temperature drops by up to 10 °C, and the average altitude is 2200 m (it is the highest continent on the planet). For this reason (and the logistics of moving personnel and goods), most Antarctic bases are near the coast.

Both inland and on the coast, strong winds (which have reached 327 km/h) depend not on the configuration of the isobars so much as on the intense and continuous loss of heat that this area of the planet suffers, sliding from the interior to the edge of the continent in a hemorrhage of cold responsible for the inhospitable environment that you face there. "Katabatic winds are Antarctica's specialty," John Goodge of the University of Minnesota-Duluth told me at a conference; "there is no stopping them, they are like sheets, rivers flowing unimpeded from the interior of the continent to the coast." This "freezer" is necessary to counteract the excess heat that originates in the tropics and the equator. Our planet is like an immense thermodynamic machine and, like any machine of this type, it must have a source of heat and a source of cold that are in equilibrium. In the Antarctic, sinking water cools and circulates toward higher latitudes to warm up. We are looking at a basic thermostat for our own survival, as what happens in Antarctica affects us all, either by the heat balance or by the increase (or decrease) in sea level due to melting or creation of ice masses. Moreover, this cold water captures part of the carbon dioxide (CO_2) produced on the planet (whether of anthropogenic or 'natural' origin).

Not everything in this remote part of the planet creates a hostile sensation. The silence and the play of light that can form halos and mirages, or even the *auroras australis* that covers the entire sky, invite us to stay. Yet where we really find life is in its vast ocean. Below the water's surface, the hostility of the 'land' contrasts with the richest and most diverse areas of the planet. There, the abundance of nutrients allows for an explosion of life from algae and other microscopic organisms, which are food for krill, the base of the food chain in the water column.

Krill are the prey of choice for fish, seals, penguins, migratory birds and whales. But the production window is short—there's only a few months, from November to the end of March, when the incident light is sufficient to fuel the primary production machine. Antarctica is *de facto* one of the eight ecozones of our planet, due to its unique cold, aridity, strong seasonality and lack of terrestrial vegetation over practically the entire continent.

Antarctica is also a paradise for myriad scientific disciplines. Meteorologists, astronomers, glaciologists, geologists, biologists, oceanographers… they all want to visit the white continent, as it is a natural laboratory of colossal size, unparalleled anywhere else on the planet. We can understand much about the capacity for evolution, rapid changes in climate or how currents work by taking a deep and heartfelt look at the forgotten continent, even if that means using costly means that are sometimes little understood by society, which craves immediate and applicable results from science and human knowledge.

As Isolated as It Is Possible to Be

The blizzard is strongly blowing the suspended snow and ice, forming a white layer that restricts vision to less than 5 m. This is not unique to the Antarctic continent; it also happens in places like the Himalayas or Greenland, but here it is frequent and persists for weeks. Air traffic becomes impossible and reaching either temporary camps or permanent bases is extremely complicated.

This gives a real sense of isolation to the few Antarctic pilgrims or quasi-permanent inhabitants who dare to come here (nobody usually stays on a base for more than 15 months).

Antarctica's isolation has been going on for many millions of years, but it was not always like this. About 110 million years ago there were ferns, dinosaurs, insects… it was a temperate system that began to drift southward, separating from Australia and South Africa (now more than 3800 km away) to distance itself from South America. The environment became colder and colder, but there was still a corridor to the 'cone' of South America until about 35 to 30 million years ago, when the link with that continent was broken. Possibly the first ice sheets originated about 34 million years ago, in line with a sharp drop in atmospheric CO_2 that caused an average cooling of about 4 °C across the planet. The movement of other continents diminished. With the formation of the Polar Front, the continent became definitively isolated from the rest of the planet, especially from a biological point of view, making it the perfect place to understand the planet's thermodynamics and, at the same time, somewhere suitable only for organisms able to resist the tremendous imposition of low temperature and strong seasonality.

In fact, Antarctic fauna seems isolated due to the large number of endemics in this area of the planet. "The Antarctic Polar Front has always been seen as a major physical barrier to the flow of organisms from north to south and vice versa," says Andrew Clarke of the British Antarctic Survey of Great Britain, "but we may be seeing changes in trends perhaps due to local changes in climate or the direct (and involuntary) action of man, who introduces species especially in the area of the Peninsula." The isolation of Antarctica is not continuous; there are places where there are exchanges of fauna, such as at the northernmost tip, barely a thousand kilometers from South America, he adds: "Perhaps the most relevant evidence is that of the decapods (crabs), whose larvae have penetrated waters considered Antarctic." This group of animals is not yet present in the rich fauna of its icy waters but currently seems to be adapting well, in some cases, to the extreme conditions of its waters.

The difficult question remains as to what caused the continent to drift. Plate tectonics *per se* does not seem to have been the sole culprit behind the ostracism of the white continent (it acts as a thermoregulator of the planet from total isolation). Ralph von Frese of Ohio State University has an interesting hypothesis about this subject that he presented at a polar congress: "We have found a crater of about 300 miles that impacted about 250 million years ago, probably causing the largest extinction on the planet (Permian–Triassic extinction)," says this American scientist. The brutal impact may have created the right conditions for tectonic plates to push Australia upward and Antarctica downward. Slowly but inexorably, the white continent would have abandoned its comfortable northern position and moved toward the southernmost part of the southern hemisphere. "There are more than twenty impacts of that magnitude on the moon," stresses von Frese; "it should not surprise us at all that one of those comets or meteorites impacted the Earth hundreds of millions of years ago." A meteorite larger than Chicxulub (the one that probably extinguished the dinosaurs) would have nearly finished life on our planet (thus giving the dinosaurs the chance to develop to their full potential), laying the groundwork for the continent to isolate itself and partly govern our climate.

2

A Brief History of an Arduous Discovery

Heroes of the Cold

Sitting on the couch or on the floor at home, my family all looked forward to watching Jacques-Yves Cousteau on Sunday afternoons before the news. It was one of the few worthwhile programs on television in the 1970s, after a binge of *Little House on the Prairie* or similar, *Fantástico*, or the usual bullfighting or other unbearable event on either of the two channels in Spain. Those 40 min of intrepid travel made us dream of worlds as remote as Greece, Papua or the Red Sea at a time when there were still no low-cost companies to take us at an affordable cost. The day came when Cousteau and his men set sail for the Antarctic continent and… they had a hard time. The voyage nearly cost them their lives (really) and was a foolhardy thing to do, as Cousteau himself would later admit. At the time, a boy with aspirations of being a biologist, like me, did not realize what was really happening on a ship that was completely unsuited to navigating the most dangerous waters on our planet, the fierce cold that bit through the installations and gears, and the relentless wind that whipped the *Calypso,* bobbing around like a nutshell. Many years later I understood what it meant to go to the white continent in the comfort of the world's most polar-ready oceanographic vessel, the *Polarstern*. Cousteau's crew did not dramatize their report; they must have had a terrible voyage, especially at that time, when it was more complicated than now to reach that latitude due to far poorer communications and a deeper sense of isolation and danger.

I now also understand the sense of beauty, bewilderment and vulnerability of James Cook who, without ever seeing the Antarctic continent, in 1773 glimpsed several islands penetrating beyond the Antarctic circumpolar

© The Author(s), under exclusive license to Springer Nature Switzerland AG 2022
S. Rossi, *A Journey in Antarctica*, Springer Praxis Books,
https://doi.org/10.1007/978-3-030-89492-4_2

current. He describes the walls of ice, the silence and the feeling of loneliness, of the end of the world. After his expeditions, everyone understood that they were facing a place that in some ways resembled the Arctic, yet few had sensed that *Terra Australis* would be a huge continent. In 1820, Fabian Gottlieb von Bellingshausen, a Russian explorer, described the Antarctic's shores of the Peninsula, its friendliest and most accessible part.

An engine of exploration had been set in motion, an eagerness to know more about this remote and hostile part of the planet. Many understood that it was one of the destinations to be explored not to make profit (until the whaling industry spread into the area, decades later) but for the glory and honor of collecting the most findings or to be the first to reach unexplored areas. John Davis was the first man officially to set foot on the continent and soon after began exploration of the Ross and Weddell Seas, the chief frozen seas of our planet.

It was not until the early twentieth century that expeditions began to be made in search of the South Pole, one of the last bastions of this untouched mainland. In the so-called Heroic Era, a time in history when the most audacious geographical discoveries were being made (personally, I always think that there is nothing more audacious than Paleolithic people conquering new frontiers, but there…), two explorers were to face each other to conquer this remote place in the middle of an ice desert: the British Robert Scott and the Norwegian Roald Amundsen. The journey was expected to be difficult, perhaps one of the most complicated and hardest tasks that could be considered at that time. And expensive. It was the equivalent, in a certain sense, of an interplanetary voyage, but this type of feat would bring glory to the country that promoted it, to the sponsors who financed it and, of course, to the men who risked their lives to go beyond what anyone had achieved at that time.

Amundsen can be regarded as perhaps the best expert polar explorer there has ever been. He began undertaking long expeditions with Gerlache around 1899, but it was not until a decade later that he secured funding for his own, to conquer the South Pole, having lost the race to the North Pole to Robert Peary. He knew that a Briton, Scott, was also mounting an expedition for the same purpose. Amundsen, who was in debt, knew that the conquest of the South Pole could bring him glory and money, a lot of money. He took 97 dogs and tens of tons of equipment to make the voyage in stages, focusing more on the conquest than on the science. The English were equipped for a more scientific expedition, preparing to carry out meteorological measurements, glacier observations and even cartography. Amundsen sent a message to his competitor, Scott, warning him of a change of route. This unsettled the Briton. Both were uneasy, because they knew that they had just a single chance to beat their rival. In both camps a strict routine was preserved, with moments of relaxation that meant, for the Norwegians, construction of a small sauna in the inner camp.

After both groups spent the winter in Antarctica, on 24 August 1911 the sun appeared again, reactivating their expeditions. Spending a winter in Antarctica today is very hard, feeling the isolation and knowing that you are trapped where it may be impossible to leave for weeks, but doing so in 1911 was a tremendous feat. There was no way of knowing what was going on outside, and they were fully aware that no one could come to their rescue in an emergency. The cold, the violent wind, the lack of sunshine and the diet limited to penguins, seals and fish as the main course… I am convinced that it was the closest thing to hell that we can imagine, surrounded by a relentless ocean that has wrecked so many ships with its terrible violence.

After disputes and hesitations, Amundsen's group arrived at 82°S on 4 November, not knowing whether Scott's group was ahead or behind them. Their uncertainty was mixed with anxiety, as the men were lacerated by the cold, the loneliness and the blizzards that allowed journeys of only about twenty miles a day, plagued by dangerous crevasses that could mean an unexpected death. On 14 December 1911 at 0300 h, Amundsen gave the command, "Halt!," to the four expedition members who were accompanying him. They had reached the South Pole. They planted the flag and made a circle, embracing each other. It was an unprecedented feat. Those men were not wearing Gore-Tex clothing, Thermolactyl underpants or special cushioned footwear. Nor did they have a satellite connection that could help them psychologically, knowing that at any moment they could call for help and have someone come to their rescue. But they knew they were the first humans to set foot on that part of the planet, and I'm convinced that this must have thrilled and stunned them.

A month later the picture was very different. After a journey filled with hardship, Scott and his men saw the Norwegian flag flying, taking away the glory of being first. Tears, despair and bewilderment must have gripped them at the terrible sight. For me, Scott had as much credit as Amundsen. He came second, yet his feat was very similar to that of the Norwegian. He was the loser for not having the glory, but a winner for having gone where only five men had gone before.

They found provisions left behind by the Norwegians, who most certainly thought of their rivals as men who were going to need all the support that they could to return. Because once you get there, you have to go back. And Scott's men did not return. They perished on the way, exhausted and terrified by the cold, cold that has been proven in recent reports to be much harsher than usual at that time.

The men who continued to work on the white continent realized one thing: in such a hostile environment, only cooperation can guarantee success. This is why men like Georg von Neumayer, a German explorer and scientist, took this path of collaboration as a means to conquer one of the most remote and inaccessible places on the planet. It was necessary to establish meteorological and scientific bases that would maintain continuous contact, something that would keep people united in the face of adversity. This would prove to be the perfect way to advance studies of the continent, yet it would not crystallize until after the two major armed conflicts that mankind has suffered during its history.

Shackleton and the Human Spirit

Until 1959 (upon constitution of the Antarctic Treaty, considered 'non-routine'), there were more than 300 official expeditions to Antarctica. We all remember two: by Amundsen and Scott (the great race for glory); and by Shackleton (the triumph of the human spirit of survival, mixed with incredible luck).

Shortly before World War I (August 1914), having read a poster reading "Men wanted for dangerous voyage. Low pay. Extreme cold. Long months of complete darkness. Constant danger. Not safe to return alive," the men departed who had succumbed to the influence of adventure and glory. Their leader, Ernest Shackleton, intended to cross the Antarctic continent with part of the 28-strong team that accompanied him. It was not the first time that Shackleton had embarked on an adventure on the white continent, nor would it be the last. In 1907 he had attempted to reach the South Pole, but 160 km away had turned back as it was too risky (leaving in his hut some cases of whiskey, later found by archaeologists). His passion for the Antarctic continent was deep and in order to reach where no other man had been he was willing to make more precarious journeys than other explorers.

Crossing 3300 km of ice today seems a major feat but, with our technology, it is feasible. At that time, crossing from the Weddell to the Ross Sea was more than a feat: it was near suicide. After penetrating the Weddell, the ship *Endurance* was trapped in Vahsel Bay. The 28 men did everything humanly possible to get the ship out of the trap, but the ice tightened its circle, impeding the small margin for movement that remained. Once the ship had been immobilized, the expanding ice began to press against the hull. In the middle of the night, the men had to climb out and take refuge on the ice. The photographer, Australian Frank Hurley, re-entered as the ship was sinking to retrieve invaluable photographic material, which has survived to this day. I can imagine him going onboard the tilted ship into which water was slowly seeping, and also those men who saw their hopes of getting out alive vanish, engulfed by one of the most unforgiving seas on the planet.

Against the emptiness that those men must have felt, Shackleton was determined to save them, no matter what. Only a strong character, perhaps one of the strongest leaders in impossible circumstances, was able to keep calm and give a small glimmer of hope to the 28 men who found themselves in the middle of nowhere, waiting only for an atrocious death. After giving precise instructions, Shackleton embarked with some of his men in a boat under 7 m long, crossing the frozen sea with a mix of precision and luck that managed to get them to Elephant Island. Once there, they crossed to the far side of the island in late autumn via a glacier and an inhospitable mountainous area that led them to a small base. Here, the people could not believe Shackleton's story of luck and determination.

After a few short days of rest they went to look for the remaining 22 men who, I am certain, were expecting only death. The faces of these men, their feelings, the light that in those critical moments must have come to the fore cannot be described. In July, in the middle of the Antarctic winter, Shackleton and all his men returned alive to Punta Arenas. The expedition had been a resounding failure, yet it showed the spirit of human endurance and true heroism.

When I am at sea, struggling in an inflatable boat because of bad weather, cold or some insignificant setback, I think of Shackleton and his men. It's a way of making you realize how ridiculous we can be in the face of problems in a world where we have everything and the real risk is minimal, compared to that in their exploits. Without any doubt, if I had to be lost in the middle of the Antarctic continent with someone, it would be with a man like Shackleton.

3

Paleoclimate: Looking for the Importance of Antarctica

Hidden Traces in the Ice and in the Ground

Less than 15 years ago, the prospect of understanding the climate millions of years ago did not seem credible to me, when we still do not understand our current climate. During the second *Polarstern* expedition (2003/2004) I met two Englishmen who were trying to understand the dynamics of the Antarctic ice floe through satellite imaging and using devices to track surface sea currents (Fig. 3.1). These were indirect measurements, and I was fascinated when they explained to me that the ice and sediments could tell stories about the climate of yesteryear. Of course, I had studied these kinds of indicators in my undergraduate studies, but I barely understood them because I found it too difficult to draw conclusions from isotopes, molecules or microfossils. Even today I find it difficult, because of my more direct and empirical training. Although sometimes I find it somewhat speculative, it must be recognized that this type of work on the studies of the Earth's climate was of titanic importance before there were direct measurement tools.

Paleoclimatic study of the Antarctic continent is essential to understand the changes throughout the planet's history and those that are occurring today. In particular, the focus of many specialists is the last million years (in which the presence and frequency of glaciation have been fundamental to the evolution of life on Earth), and they are in a frantic race against the clock to understand when, how and what repercussions the various climate transformations have had on our planet. During this period, especially the last 400,000 years, glacial and interglacial periods have increased in frequency, occurring every 10,000 years or so. However, recent studies have detected that temperature

© The Author(s), under exclusive license to Springer Nature Switzerland AG 2022
S. Rossi, *A Journey in Antarctica*, Springer Praxis Books,
https://doi.org/10.1007/978-3-030-89492-4_3

Fig. 3.1 The German ship Polarstern, the biggest oceanographic ice breaker of the world

differences of 6 °C–7 °C from our own time also occurred much earlier. For instance, in the Eocene the CO_2 concentration reached levels much higher than today (between 1000 and 4000 ppm). In this 'world without ice', the Earth adapted through major changes in sea level until about 14 million years ago. Then, in Antarctica, the isolation due to the circumpolar current was at its maximum, causing temperatures to decrease. The sea at that time had already forced organisms to adapt to the cold, allowing the survival of only those organisms that could best withstand the cold and the strong seasonality typical of high latitudes. Specific examples are the appearance in fish of anti-freeze substances and the disappearance in this type of vertebrate of hemoglobin as a gas transporter.

Climate has always been one of the main drivers of the evolution of life. It depends, in large part, on ocean currents and their heat transport. Alfredo Martínez-García of the Geological Institute of Zurich studies the dynamics of these currents millions of years ago. "We have been able to see that the sub-Arctic and sub-Antarctic regions are cooling almost simultaneously due to a tongue of cold water," says the scientist. The changes are mainly due to movements of the Earth's orbit that transform the currents from subtropical to subpolar patterns. These are accelerated changes mainly at the poles, although in the Arctic faster variations (such as occur now) are detected than in Antarctica. As Antarctica has become more isolated, the climatic changes have differed between north and south. In the Holocene, for example (the period that includes the last interglacial period, i.e. from 12,000 to 10,000 years until now), Antarctica began to thaw before the immense glacial masses of the boreal hemisphere. "The changes accelerated mostly about 8000–5000 years ago, and about 4700–2000 years ago there was a period that we could consider warm on the white continent," commented Michael Prentice of the University of New Hampshire in the United States at a conference.

Temperature increases in this and other places on the planet are always linked to CO_2 fluctuations. In cold periods, CO_2 concentrations can be as much as 80 parts per million (ppm) lower than today. "Changes in planetary climatology are anything but slow," says Richard Alley of the State University of Pennsylvania (United States). For example, during the last ice age the climate underwent a series of abrupt changes: in just a few decades, temperatures in Greenland rose by more than 15 °C. Once again, Antarctica is behind this. "In periods of low CO_2 values," explains Britton Stephens of the Scripps Institution of Oceanography in California, "it may be that the ventilation of the deep waters associated with Antarctica is poor because of the extensive ice cover, lowering the exchange of gases where the 'sinks' for this type of gas are located." What Stephens and other scientists are beginning to suspect is that there is no warning in advance of these abrupt changes. We are yet to understand much of the dynamics, how the change occurs and what affects what. We have evidence that things have happened, even how and when they happened, but we get lost when we try to understand the exact mechanism and what causes it.

But how on Earth do scientists come to these conclusions? The common tool is proxy indicators. Several parameters are used simultaneously, such as the magnesium to calcium ratio, oxygen isotope concentrations, long-lived organic molecules (lipids) such as alkenones and certain organisms that we know are able to withstand cold or heat. The marine sediment is drilled hundreds or thousands of meters deep and divided into slices measuring millimeters to centimeters thick. These are studied in laboratories with exquisitely delicate care to avoid contamination. From the 218 oceanographic expeditions of the Ocean Drilling program (which began in 1969, the same year that man set foot on the Moon), 332,370 m of sediment samples have now been collected, allowing in-depth studies in various parts of the world (including Antarctica) into the changes in our planet's climate. "The proper treatment of the material is essential to avoid making mistakes," says Robert Burger of Yale University (United States).

Ice, however, is an even better climate witness than sediment, because no organisms have disturbed, punctured or eaten it. Bubbles of air become trapped within ice and their concentration of gases of all kinds can be analyzed even after thousands of years. In Antarctica, the ice can be more than 3000 meters thick in certain parts of the continent, so we can trace the different glacial and interglacial periods over more than a hundred thousand years.

Who Was in Antarctica Before Our Time?

As mentioned, Antarctica was not always like this. In fact, it is one of the areas of our planet that has undergone the most change, cornering and extinguishing many species unable to withstand its cold. In the Permian (about 300–250 million years ago), forests harbored life that was much influenced by strong seasonality. "These were deciduous forests," says Edith Taylor of Ohio State University (United States), "because of the lack of light for a considerable time." An evergreen forest would have forced respiration in the trees by having to maintain the living structure of the cells capable of photosynthesizing, and this could have caused excessive stress. Not everyone agrees; it is known that there are conifers today capable of withstanding long periods without light, but it is clear that temperatures of –30 or –40 °C could be reached in the middle of winter in certain areas of the still-annexed Antarctic continent. "I don't believe that temperatures were so severe throughout the continent," adds Taylor; "although there is geological and climatic evidence of the area, we must always take into account the biology of the species and their limits of tolerance."

That is why Michael Benton of the University of Bristol (Great Britain) wonders how the dinosaurs whose fossils have been found both in Antarctica and near the Arctic were able to live tens of millions of years later. "The dinosaurs that lived here and for which we have a fossil record were in places where, of course, it was cold and there was strong seasonality," comments Benton. Dinosaur remains and other reptiles (ornithopods and ceratopsids, turtles and crocodiles) from 114 to 66 million years ago have been found well above 60° latitude (even between 75° and 85°, both north and south). Calculations through indicators show intense cold in these areas and large expanses of ice, a very different picture from the swampy, warm, vegetation-filled areas typical of dinosaur movies. "It's impossible that they were not warm-blooded," Benton reflects, "or they could not have survived at such low temperatures."

But there are scientists who doubt that there were large expanses of ice at that time, even at such high latitudes. If temperatures had remained between 5 °C and 13 °C (between 0 °C and 5 °C in Antarctica), it is possible that they would not have needed thermal regulation (warm blood) as we understand it today. In fact, some specialists have resorted to gigantothermy to explain their presence: large animals would have been able to maintain a warmer core inside due to their proportions. "What is clear," says Benton, "is that if the explanation is that they were able to maintain a constant temperature thanks to their activity and their large volume, the amount of food they needed was immense." Hibernation seems impossible because of their size, and migration implausible. How could they move 3700–4600 km south each year in search of warmer lands?

Dinosaurs at the poles create a key dilemma for their physiology and ecology. It is an issue that is far from being resolved and it serves as both an argument and counter-argument between groups of specialists, because what is clear is that these fascinating animals lived in a period when there was little warmth there. Once the dinosaurs disappeared, mammals took over. Primitive marsupials and reptiles moved among large ferns and giant trees, living very well in the Eocene, a time (as we have explained above) when temperatures were very high compared to the present.

And at the bottom of the sea, what was going on? A recent hypothesis developed by Josep Maria Gili of the Institute of Marine Sciences of Barcelona (CSIC) places the benthic system of the Ross Sea and the Weddell Sea (Antarctica) as a relic of the organisms that developed in the Cretaceous in parallel to the dinosaurs. In these southern bottoms, the suspensivores (sponges, gorgonians and bryozoans) live on soft substrate, forming dense patches whose variety of species and biomass rivals even those of coral reefs. It is a unique system, because in the continental shelves of the rest of the planet these suspensivores are only well represented on hard bottoms (rocks, walls, atolls); soft bottoms clog the fragile structures where sponges, gorgonians or bryozoans feed because of the large amount of suspended matter from rivers or detritus from underwater plants. In Antarctica there are no major rivers, and only in the Peninsula are there large, submerged algae. The Antarctic benthic system is therefore peculiar, a direct descendant of the great Cretaceous cataclysm due to conditions unaltered since then in this part of the planet.

At that time, the meteorite impact would have created a dense dust cloud, causing the collapse of the oceanic primary production by blocking sunlight penetration for a long time, which would have prevented photosynthesis in microscopic algae, the food base of the entire oceanic system of the planet. Yet the organisms living around Antarctica were already accustomed to the twin effects that the great cataclysm would have caused: a prolonged lack of food due to extended periods without light; and a significant drop in temperature, which also affected the phytoplanktonic machinery greatly. The organisms found on the Antarctic seabed would have survived the cataclysm by their life cycle's adaptation to lack of light and low productivity for months at a time. Indeed, Antarctic gorgonians and sponges are considered today to be the world's most primitive groups, and over millions of years they will most likely radiate to the rest of the planet. The white continent, a fragile ecosystem increasingly disturbed by the hand of man, once again gives us an impressive evolutionary lesson of how nature is capable of surviving the most adverse conditions, even in the face of one of the most serious planetary crises. To think that some groups of cnidarians or porifera could have remained almost unchanged for such a long time gives me a strange feeling.

But how did the organisms survive that had to endure the intense glaciations of the past million years? The situation was quite different from today and, of course, saw extremes as in the Eocene. For example, during the last glacial period the conditions in Antarctica were even harsher than today. Surface fauna lived only in very specific areas, the polynyas. These are areas devoid of multiannual ice, thanks to the wind and the action of sea currents. They are the result of the push of katabatic winds toward the open sea and the upwelling of water due to the presence of submarine mountains forcing currents to rise to the surface. "We think they were survival strongholds," says Sven Thatje of the National Oceanography Centre in Southampton (Great Britain); "polynyas were essential for birds, seals and even whales in this part of the planet, and they must not have been very abundant." The extent of the sea ice was much greater and was devoid of seasonality over much larger areas than now. Productivity was therefore restricted to those places where the light reached, allowing phytoplankton organisms (microscopic algae) to take advantage of nutrients and grow. Few were able to withstand such harsh conditions for survival, and the populations must have been much smaller than today, where the summer bonanza is strongly marked and allows even giants like whales to feed on the large amount of food from the growth of phytoplankton once the sea ice disappears each year.

Let's Not Forget the Eocene

Let's spend a couple of minutes on a period in the history of our planet in which things happened that we do not yet understand yet whose effects revolutionized life: the Eocene. This period, which spanned 56 million years (ma) to 48.6 ma, was brief but intense. Over about a thousand years an abrupt warming occurred that lasted about 150,000 to 170,000 years, raising temperatures on average by 5 °C–7 °C across all the planet. As already mentioned, this was due to an anomaly in CO_2 concentrations, which reached 1000–4000 ppm (today we are heading toward 415 ppm and are already very worried), raising sea level by an average of 70 m, thawing the entire planet and increasing the acidification of the oceans to levels that caused the extinction of many creatures adapted to marine waters with a higher pH. In the polar regions, temperatures also rose, in some areas to near those of subtropical climates, with waters of more than 12 °C (today they are between 0 °C and -2 °C), and this drastically changed the flora and fauna. There was a great extinction, yet some organisms took advantage to create new evolutionary routes and prosper. We ourselves are partly the result of these changes, because mammals definitively took over from the now-extinct dinosaurs, creating a diversity that had not existed until then in this animal group.

It is not known how so much CO_2 arose all at once—what we do know is that it took more than 150,000 years to be reabsorbed. The high level could have been achieved by methane hydrates on the ocean floor and in the tundra being in unstable equilibrium, or by a series of volcanic eruptions suddenly injecting this gas that has sustained life on our planet, maintaining temperatures conducive to life as we know it here on Earth. We do not know. Such an injection in such a short time must have left much of the biota unable to react, but those species that could certainly took the opportunity to expand. Later, in the Cenozoic, the opposite conditions were created; that is, the climatological steps toward glaciation, as repeated many times in the planet's evolutionary panorama. As already mentioned, this phenomenon caused the formation of Antarctica's permanent ice sheet (about 35–30 million years ago), a path from which, due to its isolation from the rest of the planet, this region had no return.

The Eocene remains a mystery because, as in other glacial and interglacial epochs, what triggered this large-scale climatic phenomenon remains an enigma. However, we do know one thing: its consequences. And we are accelerating toward the same place. If we continue with our current trend of gas emissions, the most conservative models talk about exceeding 1000 ppm CO_2 in about a hundred years. Can we wait and see? In my opinion, we are so stupid that we shan't do anything about it. We're all right for one more day, aren't we?

4

Survival Lessons: Mars on Earth

Studying Life in Space Without Leaving Earth

I well remember Carl Sagan's documentaries in which he explained the exploration to be carried out on Mars to find out (among many other things) if there could be life on that planet. The spacecraft landed, an unmanned vehicle came out, cruised around a specific area and carried out a series of 'tastings' to find out whether or not there could be some kind of life on the ground. What I did not know at that time (early 1980s) was that men and women from several universities and institutions had been working for years on how to address this problem through various experiments on the white continent.

Dry Valley in Antarctica is considered one of the most inhospitable places on our planet. It has no ice, and the precipitation (in the form of snow) does not usually exceed 15 grams per square centimeter. The katabatic winds (from land to sea) bring in an unbearable cold, making average temperatures throughout the year of below -20 °C to -25 °C. The little snow that falls immediately sublimates—it does not even have time to freeze. That is why already by the 1970s several scientific groups had identified this place as one of the most suitable for research on extreme situations for life. Rather than frightening scientists, Robert Falcon Scott's description of Dry Valley as "the valley of death" stimulated a type of science in which answers were sought for environments 'impossible' for life, in which the lack of water and extreme temperatures put any organism that tried to colonize it to the test.

"The metabolism of organisms in such a place must be very, very slow," says Dr. Horowitz of the Pasadena Institute of Technology in California. However, those same organisms must also be extremely efficient. Nothing can escape

© The Author(s), under exclusive license to Springer Nature Switzerland AG 2022
S. Rossi, *A Journey in Antarctica*, Springer Praxis Books,
https://doi.org/10.1007/978-3-030-89492-4_4

them, because 'nothing' is what they have. In such a place, where the surface of the lakes is permanently frozen, the overwhelming majority of life is microbial. It is no place for metazoans or metaphytes, no place for plants or animals. "The very wind that whips the organisms is possibly the vector necessary for their dispersal," Horowitz continues. "There are a large number of endemic diatom species," observes Joshua Darling of the University of Colorado; "this is because only a few highly adapted ones are able to survive in such a harsh environment." Slight shifts result in the dominance of one species or another in a game of extreme survival. In a place where life is the exception, the conditions for the study for exobiology or astrobiology are ideal.

Certainly in such a hostile environment, where the density of organisms is so low, it is essential to ensure that your methodology won't let you down. Scientist Brajesh Singh of the Macaulay Research Institute, Great Britain, admires recent work done on bacterial diversity in an area as hostile as Dry Valley; however, he doubts whether absolutely everything has been done to extract every last vestige of life, because it is a highly complicated matter to take samples on Mars or any another planet or moon. You only get a single chance. A crucial group of organisms found in other extreme places, like in the Chilean high mountain lakes or alpine glaciers, namely bacteria called archaea, were not found to be present in that study "perhaps because of very low densities or inappropriate extractions". In places like Dry Valley you can have another chance, repeating the experiment, but what about Mars? What about Jupiter's satellite, Europa, or the very distant Pluto?

This is a good question. The experimental design has to be perfect. It cannot have any blemishes and, above all, it must not contaminate the sample or the site. And that's a very difficult thing. Well, basically, it's impossible. Just ask the various committees involved in accessing one of the most isolated places on Earth, Lake Vostok, without contaminating it.

Under more than 3 km of ice (in some points, 4 km), Lake Vostok is one of the largest lakes in Antarctica. With a surface area of about 14,000 square kilometers, more than 1800 cubic kilometers of ice, 5400 cubic kilometers of water and a maximum depth of about 1067 m, it is a dormant lake, undisturbed for the past 14–15 million years. The water remains liquid despite its low temperature (-3 °C) due to the enormous pressures (about 380 atmospheres). "It's a fascinating place," says Talalay Pavel of China's Jilin University; "it took more than 22 years of exploration to get to the bottom to collect samples by the hard-nosed scientists who have been working there for a long time."

In this place, which seems to me to be one of the darkest places on the planet, there is life, but at very low concentrations (no more than 200–300 bacteria per cubic millimeter, when in alpine lakes with practically no nutrients there's about 10–20 million per cubic millimeter), mainly due to its lack of 'food'. But there is life. What is not clear, however, is its source of energy.

Light is absent, so it must be a system based on chemistry, transforming inorganic molecules into usable energy (chemolithotrophy). The glaciers and their slow, small contribution could be the basis for the nutrients, but there is also speculation about an active geothermal source that would propitiate the complex circulation of the lake's deep abysses. Toward the end of the 1990s, Dr. Jouzel and his colleagues at the Laboratoire des Sciences du Climat et de l'Environment of the France's National Centre for Scientific Research (CNR) began to make calculations using equations based on stable isotopes, demonstrating that the Vostok was an open system with water renewed approximately every 420,000 years; water does flow in and out, yet at an extremely slow rate. Under ice that may be 420,000 years old, this research groups and others were working at the limits, aware that penetrating the final phase implied perturbing the system in a surely irreversible way.

"The jewel in the crown is to reach the waters of Vostok; so far we have positive indications of life from interstitial water accumulated in the ice itself," says Cyan Ellis-Evans of the British Antarctic Survey. This scientist repeats how delicate a task it is to 'touch' the water of this and other lakes beneath the deep ice of the white continent. "Completely sterile material must be used, but we are aware that 'zero contamination' is impossible," adds Ellis-Evans in the Washington Post. Had biologists not intervened, it is more than likely that this lake would have been contaminated forever, polluted by geologists' drills. The environment that had remained undisturbed for millions of years would suddenly be 'infected' by bacteria and other organisms from the outside. That's not all; these sub-glacial lakes are known to be connected. There could easily be cross-contagion, wrecking millions and millions of years of isolation.

Reaching the lake without any contamination is impossible. The question is what represents the tolerable limit of such contamination. Let's proceed on the basis that what may seem ridiculous to us, a few tens or hundreds of bacteria per cubic millimeter, represents a total change in such a pristine system.

Why is this research so important? Just think about the ultimate goal: to create the sophisticated tools to explore satellites like Europa or planets like Mars. "The drilling project is especially important for these environments," Jouzel stresses. "Exobiology or the search for life on other planets is already being tested on our planet before visiting the next ones." In any case, the race to the treasure is blinding many scientists to the potential contamination. Although the Antarctic and Southern Ocean Coalition was strongly opposed any kind of interference in Lake Vostok, the interests of the main protagonists and the desire to be in the press almost broke its will and its means of preventing the fragmentation of the last few meters separating the drill bits from the icy lake water that had remained undisturbed for so many millions of years.

When the researchers finally reached the lake, they began to study the samples meticulously. Life was found, very sparse yet with considerable biodiversity by being so poor. "These lakes may be suitable for prokaryotes, such as bacteria, but hostile to complex organisms such as metazoans," explains Sven Thatje of the National Oceanography Centre at the University of Southampton; "however, their extreme conditions invite us to explore places like Jupiter's moon Europa, which has potential geothermal within it." Exobiology puts much effort into perfecting sample collection methods at places like Dry Valley or Lake Vostok, because if you visit a moon of Saturn you don't want to contaminate something billions of years old that has remained undisturbed until now.

Changes in the Imperturbable

Even in such a remote area of the planet as Dry Valley, it appears that certain changes could become a reality over the next few decades. "The accumulation of snow in this remote desert area may occur over a short period of time," explains Eduard Ayres of Colorado State University. In such a dry and cold place, the presence of snow could be the key to understanding future changes. Although in the experimental set-up of Ayres and his colleagues very few chemical characteristics of the soil altered upon the introduction of snow, the presence of soil moisture could favor one or other organism. One that had hitherto been unsuccessful and remained 'dormant' could take over from one that was disadvantaged by the change. This is what Dr. Andriuzzi of Colorado State University has been able to reveal after more than 20 years of monitoring: "there is more humidity than before, more snow, and some dominant organisms are giving way to others more adapted to a life, let's say, 'less hard'".

It should be noted that, in general, those species that resist extreme conditions are very poor competitors when a bonanza arrives in the ecosystem. Perhaps that is why, when a very localized snowmelt between 2001 and 2002 wetted part of Dry Valley, some nematode worms were favored and proliferated, displacing another group more accustomed to a drier and colder climate. Far from seeing this disturbance as something negative and the result of ever-present climate change, Dr. Banet and his collaborators at Virginia Polytechnic University in the United States see it as a reflection of what is happening in other desert areas of the planet. "What happens is that in this place these changes are even more sporadic," adds the American scientist. "At a given moment, species that have not played any relevant role in the ecosystem take up positions in the face of these changes that occur from time to time and are only detected thanks to the long time series." The same trend of change is observed by Thomas Niederberger and his collaborators at the University of Delaware (United States) in microorganisms such as bacteria: "microbial activity is soaring, because there is more moisture due to climate change," he explains. It seems incredible that a place so remote and with such a small number of species can remind us that biodiversity is important (among other things): the existence of several species is what keeps any system alive, dynamic and flexible. All these communities are highly sensitive to climate change, activity accelerates and biogeochemical changes cause transformations. Even in the most inhospitable place on Earth.

5

Perpetually Moving Glaciers: Where Icebergs Emerge

Glaciers, Source of Titans

I like the Scandinavian legend that tells how, contrary to what southern cultures claim, the end of the world will be caused not by fire but by ice, formed in endless winters that will freeze even the seas. We always imagine flames and hardship from scorching heat, and it is hard for us to see how difficult is life surrounded by a desert of ice like in Antarctica. The first time I saw a cracked glacier from the air in 2000 I was shocked to see the precariousness of the ice hanging in the air, waiting to be pushed by the solid (and liquid) river, stretching miles and miles behind it.

Science was slow to study ice and its properties, especially those of glaciers. The study of ice became serious when in 1837 a young scientist, Louis Agassiz, declared before the Swiss Natural History Society the importance of glaciers and ice in shaping the morphology and rock composition of valleys and mountains: "I am going to tell you about glaciers, moraines and erratic blocks." His speech at first perplexed the audience; however, the conference participants soon understood that ice was responsible for certain geological formations that would be incomprehensible without taking glacial dynamics into account.

It is a mistake to view ice as a simple accumulation of solidified water. Ice is a large reservoir of liquid in solid state, and is an important component of our planet's global energy balance and its thermal equilibrium. Its only 'permanent' ice is continental ice, which can reach a thickness of more than 3 km on the white continent. The rest is the sea ice, which fluctuates with the heat and cold of the seasons, as we have seen. We can then understand the

© The Author(s), under exclusive license to Springer Nature Switzerland AG 2022
S. Rossi, *A Journey in Antarctica*, Springer Praxis Books,
https://doi.org/10.1007/978-3-030-89492-4_5

importance of this physical state of water, which exchanges a large amount of heat as it freezes. In addition, because the surface of the poles is white, the reflection of sunlight (mirror effect) greatly influences the balance of radiation that reaches our planet. In Antarctica, the icecap or *indlandsis* is 22 million square kilometers and represents as much as 70–75% of our planet's fresh water. If it were to melt completely, sea level would rise by 45–60 m.

It is essential to understand what is happening in this more permanent part of the white continent that is largely responsible for the climate of our entire planet. Nobody doubts anymore that glaciers all over the planet are suffering accelerated regression, but it is a complex subject in which strong asymmetries between areas are detected and in which specialists are racing to try to compare their results and intercalibrate their methods. Efforts to understand the phenomenon and its speed are increasing, because in areas where it was thought that ice was increasing it is also decreasing, and in areas where data were collected to explained the erosion speeds it is acknowledged that these are much lower than are actually observed. Even top specialists do not fully understand the dynamics of glaciers. The progression of these icy snakes requires highly complex mathematical models to understand the mechanisms, velocities and trends, and these tools are not yet perfected.

"Why is it so difficult to control the movements of glaciers?" asks David Vaughan of the British Antarctic Survey; "it is assumed that they follow a regular motion, explained by physical laws explained in the nineteenth century and that complexities such as the Coriolis force, turbulence or the laws of inertia should have no influence on their dynamics." The problem of melting glaciers and its acceleration is not only the warmth of the air itself, with temperatures that the ice could withstand for long periods of time; it is the warm water eroding entire structures from both below and above, creating fragmentation by abruptly breaking off large chunks. The bottom is not homogeneous, nor are the velocities at which areas melt. And the errors made in recording the initial data add to the confusion in constructing a sound database for the mathematical models that we mentioned earlier (Fig. 5.1).

Natural phenomena also impact on glaciers in Antarctica, adding to the complexity of interpretating the results. One observed in March 2011 was the impact of the tsunami that swept the east coast of Japan (Fukushima). Having traveled more than 13,000 km, the wave of barely 30 cm high arrived with sufficient force still to dislodge an iceberg of about 125 km² that rose 80 m from the sea's surface. Glaciers themselves create earthquakes, with a much slower speed than those felt in other parts of the world yet with a similar magnitude. Seismic waves have been detected that have been caused by glacier movements, easily moving the surroundings with an intensity of over 7 on the

Fig. 5.1 The sea is continuously interacting with the ice edge, eroding and fragmenting the shelf

Richter scale. The difference is that, while an earthquake of this magnitude usually lasts about 10 s, a glacier 100 km long and 800 m thick can cause one that lasts about 10 min, distributing the violence of the movement much more smoothly over a longer time.

Within the glacier itself, in nooks and crannies exposed to the air (which can make up 0.1–10% of its surface) in which there is liquid water, a community of microorganisms lives that has considerable influence on the capture of carbon from the atmosphere. These organisms are activated at less than 0.1 °C and they can capture a large amount: it has been calculated that across the entire continent, through photosynthesis, they are capable of retaining some 64 gigatonnes (GT) of carbon (this is the difference between what they actually produce, some 98 GT, and what they breathe, some 34 GT). This is an important figure, which proves once again that even in the most inhospitable places there is life and production. Under permanent ice in marine parts, which can extend for many kilometers, there is also life, but it is much poorer than those areas governed by seasonal ice. Mats of bacteria feed on a few isopods and amphipods, and some clueless holothurians or fish may eat these passing tenants.

Glacier Erosion

Across the planet more than 95% of so-called continental glaciers are in regression, and some have almost disappeared. The theory has been that those glaciers are the ones that responded most quickly to temperature changes, disappearing in just a few decades, while larger glaciers (such as those of Greenland or Antarctica) were more resilient, lasting for millennia. One major glacier, such as the Margerie Glacier in Glacier Bay in the northern United States, has seen its ice mass erode by some 15 km in just 20 years, and this is seen as relatively normal for a continental glacier. But this perception of quasi-immutability in imposing masses such as the Antarctic or Arctic has been proven to be false. It is true that changes are detected earlier in small glaciers, but large glaciers share the same fate. As I said before, we still do not understand well how they work. And the Intergovernmental Panel on Climate Change (IPCC) experts are nervous, because such data are basic to good predictions of rises in sea level, among other things. Only empirical data (a rise of 1.8 mm/year on average, over the past few decades) are totally reliable, and the predictions are still tentative. And, for those thinking that they will surely get better, I have some bad news: the data indicate that predictions will worsen because, as we will see later, the melting is accelerating.

"It is clear to us that one of the main problems is that the glacier calotte [icesheet] in contact with the sea may be shrinking," explains Ronja Reese of the Potsdam Institute for Climate Impact Research. Ice shelves act as a barrier to these immense masses of ice that are constantly pushing, but they may eventually crack. "In early 2017, an iceberg whose surface area was ten times the size of Manhattan Island broke off from Larsen C," continues Reese, "and at Larsen B, after the collapse before 2002, we have gone from a glacier velocity of 2.7 km^3 to more than 25 km^3." The fact that this containment barrier is getting narrower is causing glaciers to retreat inland faster than had previously been thought.

In the western part of Antarctica, almost opposite South America, many glaciers are clearly retreating. Out of 240 glaciers, 216 are in regression, or about 90%. The problem is that satellite monitoring, which is very reliable, began only in 1993, so the time series is short. Paleoclimatic data are used to reinforce the results. "Sediments and ice cores tell us that in interglacial times the sea level could have been as much as 5–9 m above the present; those observations, coupled with our satellite knowledge, in situ observations and the models are helping us understand something crucial: the rate of displacement and collapse," says Eric Steig of the University of Washington (United States).

What has been observed over the past decade is that atmospheric and oceanic conditions are changing dramatically in this area. Here, specialists are holding their breath about the future of the Thwaites Glacier. This immense mass of ice seems to have become active, accelerating its disintegration. If this is the case, in the long term there could result in a sea level rise of more than 3 m from this immense mass of ice alone.

Glaciers and their dynamics must be understood as part of the terrestrial system, yet also of the oceanic system. Fresh water flowing from glaciers in Antarctica is of great importance to the pelagic and benthic communities of the Southern Ocean. Its influence can extend 100 km, in some cases. Apart from the hydrography of the area (affected by fresh and warmer waters), there is a change to the phytoplankton and crustacean plankton, adapting to the large amount of turbidity from the meltwater, which carries sediments and muds from the land. These same muds greatly influence the benthic fauna and flora (if any), which see their filtering structures collapse, clogged by a fine suspended sludge that ends up burying many. The quality of the food may also change, with a greater amount of particles being carried from land with less nutritional value. It is therefore important to understand how they work in order to understand whether the acceleration of their melting in certain places will influence the marine fauna and flora.

The collapse of the ice shelves is therefore causing an ecosystem change. "We need to understand how these changes affect biodiversity," says Sophie Cauvy-Fraunié of the Centre de Lyon-Villeurbanne; "it seems that when the glacier retreats there is an increase in biodiversity, but sometimes it is simply because there is a transition to a new system, a transition exploited by generalists and opportunists; those who had specialized in the pre-melt dynamics disappear or are pushed into a corner." Cauvy-Fraunié has studied all types of glaciers, observing that the pattern is very similar in mountains and in fjords, at the poles and in the Himalayas. It remains to be understood what will happen in the near future because, like the rest of the planet, this situation is part of a very rapid transition, unprecedented in terms of speed rather than because it has not occurred in the past.

The Giants of the Sea

The first time I saw an iceberg I was most impressed. It was a small one, drifting far from the Antarctic continent. We came across it on the voyage from South Africa, and it looked huge to me. It was a whitish block like a house, with the rays of twilight lending it a beautiful orange tone. The veterans on that 2000 *Polarstern* expedition told me that it was not first-class, being small and colorless, so just a 'run of the mill' iceberg. Far from being disappointed, I stayed looking at it with fascination until it disappeared astern. Later I watched icebergs in all colors, textures and sizes until I was bored, but I have to admit that seeing a 'live' iceberg creates a strange feeling of smallness and at the same time of admiration that cannot be transmitted by National Geographic or Discovery Channel documentaries. No wonder that in certain Canadian towns people pay to sit on a restaurant terrace to watch icebergs while they eat, or that people travel to Perito Moreno in Argentina to see discrete icebergs floating away from the glacier when they break off.

The colors of icebergs tell us in part about the origin and longevity of the ice. Gray icebergs may denote origin in a volcanic area, while an intense blue shows the great age of the ice, having expelled under immense pressure the air inside (which makes the ice filter out all colors apart from blue). I have seen several blue icebergs, and one in particular caught my attention as it took the shape of a dragon fighting against the swell in the middle of a strong storm that tried to bring it down. One day I had the privilege of seeing something exceptional. On rare occasions during Antarctic cruises, a green iceberg crosses scientists' path. The emerald color makes it almost like the precious gemstone, with shades that sweep through a wide spectrum, depending on the inclination and penetration of light. We stayed for a long time on the various decks of the ship to observe the spectacle, while the second officer took delight in steering around it for our enjoyment. Now shrinking, the (once) green giant churned, licked by the swell at the base in slow but sure extinction. It was an old iceberg, about to be consumed by the relentless constancy of the sea, which erodes each and every iceberg, little by little.

Some people who have had the opportunity to study these green icebergs have found iron, copper and other metallic components. When young, an iceberg crawls across the sea bottom, picking up particles that are incorporated into its interior. It is unclear how this process occurs. Gradually, the iceberg is worn away and, like them all, eventually turns upside down. Then its belly is revealed, in which these metals are included. And they are not the only thing found, as microscopic organisms also seem to contribute to the color: diatoms and benthic foraminifera, small fibers… another possibility is the presence of parallel layers of green feldspars. Strong bottom currents and resuspension of the material at different times may facilitate the incorporation of these elements, which would also explain why the layers are not uniform and homogeneous. The dragging of icebergs and subsequent freezing due to changes in temperature and pressure favor the incrustation of these elements.

This sighting of a green iceberg was one of the most unexpected events recorded on the 2000 expedition. In 2004 we had the opportunity to see another, much larger and with a strange shape like a huge chair, just when we were celebrating the end of the expedition. It was another magical, unforgettable moment in which the feeling of adventure was mixed with that of scientific curiosity. We were all aware that this iceberg had little life left, since even the largest do not usually last more than 10 years due to incessant erosion by the sea.

The Effects of an Iceberg

The immense masses of ice floating aimlessly in the polar oceans have always fascinated and frightened human beings. Born of glaciers and ice floes, icebergs (ice shelves) break off due to erosion by water under the ice mass, fragmenting it and casting it adrift. They are the result of the collapses from glaciers that we talked about. Once in the sea, an iceberg is doomed to disappear in a fairly short period: about 2–3 years, ten at most if the block of ice is large and trapped in an area where temperatures are low and the erosive action of the sea is not severe. Iceberg dimensions range from a few tens of meters to kilometers, but the mass is unstable and tends to fragment easily. In November 2011, in the West Antarctic zone, a breach of more than 30 km long and 60 m deep opened up in the ice, giving rise to an floe of about 880 km^2. It emerged from Pine Island, from where it is estimated 10% of icebergs off the West Antarctic coast emerge. More than two decades earlier, an iceberg of about 5400 km^2 traveled more than 2000 km in nearly 2 years, reaching speeds of more than 13 km a day. A-68 left Larsen C in 2017 with an area of about 5800 km^2, and over a billion cubic tons of ice. It is increasingly common to see these mega-icebergs breaking away from the continent, and their influence all around is enormous. After they break apart, millions of tons of water are transferred from the solid to the liquid state. Like others before it, an iceberg will drift away from the continent to be trapped in coastal currents or go further, sometimes even beyond the circumpolar current. Icebergs of a similar size could even divide specific locations that penguin and seal populations inhabit, creating natural barriers that cause distancing between populations and a shortage of food because these creatures cannot reach their feeding areas.

As long as the ice mass floats, it will form a platform for seals and penguins on its surface and represent a source of life below (microscopic algae, small crustaceans, etc.). Its melting favors primary production. It has been proven that a large amount of life is concentrated around icebergs in the form of phytoplankton and small crustaceans, by virtue of the circulation of nutrients and micronutrients such as iron, the iceberg acting as a source of life. Its influence can extend beyond 5 km, and the large numbers floating around the continent may have an effect over no less than 39% of the Southern Ocean. "Icebergs may be favoring productivity to such an extent that in some areas phytoplankton blooms are growing larger than under normal conditions," explains David Barnes of the British Antarctic Survey. "Icebergs can fertilize the sea, causing algae to grow and fix carbon; this carbon then passes into the structures of organisms that eat them, becoming immobilized for decades or much longer." A giant iceberg, according to Barnes, may capture up to 1.9 tonnes of carbon a year, the equivalent of 400,000 cars. "That's if they don't hit the bottom… in that case we have a very destructive effect," Barnes clarifies.

In fact, in many cases an iceberg is a source of destruction, as its lower parts literally scrape the sea bottom, destroying the communities there (sponges, bryozoans, gorgonians, etc.). This phenomenon is called "iceberg scouring" and it is estimated that between 15% and 20% of the ocean floor's surface is affected (in so-called 'iceberg resting places', where icebergs are trapped, more than 50% of the area may be affected). In 2003, I witnessed the BENDEX experiment in which a 100,000 m² area of sea bottom was destroyed with special nets in an attempt to understand the effects of the destruction of an iceberg and, above all, the colonization patterns following the destruction. Over several weeks an artificial bald spot was created where the sessile fauna (sponges, gorgonians, bryozoans, ascidians, etc.) was reduced by an order of magnitude; the sediment properties remained unchanged, because it was impossible to emulate the 6–9 m deep trail (undercut) that an iceberg leaves in its wake. After 7 years (in 2011) we returned, and the first images showed that little had been restored: recovery is slow on the Antarctic seabed. In the Weddell Sea, only 5% of the area is affected by this destruction, but the number of icebergs that are breaking off may be rising, so the proportion will also increase.

Icebergs are considered one of nature's most devastating impacts on the environment, along with earthquakes, hurricanes, major fires and volcanic eruptions. The Antarctic fauna is structured by these disturbances, creating uneven patches, preferential zones and greater or lesser population densities, depending on the effect of these large masses of ice. It can take more than 250 years to recover, in a complex, stately succession (Fig. 5.2). The tracks of an iceberg, sometimes more than 140 m wide, penetrate to great depths in the sediment: we have to remember that when we see an iceberg, approximately

Fig. 5.2 A giant pycnogonid, the sea spider living in the Antarctic benthos

one-ninth is above the surface and the remaining eight parts are submerged. "In coastal areas up to 30% of the fauna and flora, especially at shallow depths, is disturbed by icebergs every year," comments Barnes; "in 2009 for example there was in a single year a disproportionate massacre of organisms on the bottom by the action of several icebergs in a coastal area of the Antarctic Peninsula." An iceberg 80 m high above the surface (very large) can have an influence at depths greater than 600 m. Icebergs are usually only a few tens of meters high, but some gigantic ones, such as those mentioned above, can even divert local ocean currents.

Formation of a Single Crystal

In a book that talks about the frozen continent, I cannot fail to dedicate a few lines to ice itself. The fundamental property of water is undoubtedly the bond that its molecules form between themselves, the hydrogen bond. This is weak, but in the predominant atmospheric conditions of planet Earth it is sufficiently robust to allow this inorganic compound to be found in three phases—solid, liquid and gas. The tetrahedron formed by the water molecule (whose center is the oxygen atom and apices the two hydrogen atoms and the electrons of hydrogen and the oxygen itself) is capable of linking to inorganic structures one after the other in a flexible yet time resistant way. While in other elements the density of molecules or atoms increases until they solidify as the temperature drops, water behaves like a rebel: it acquires its maximum density at 4 °C without having solidified, then becomes disordered again as the temperature continues to drop until its molecules fit together. This causes a curious effect that is essential to understanding life as we know it: the solid state of water, being less dense than the liquid, actually floats.

This property has allowed the permanence of aquatic life on the planet even during glacial periods, when ice banks cover seas and lakes, preventing the total solidification that would have killed all living beings there. The force that water can exert during the changes of phase (solid–liquid–gaseous) is tremendous. In the case of ice, granitic surfaces can be fragmented and wide valleys carved by its movement and weight, both of which have occurred throughout the history of our planet, shaping many of the landscapes that we currently see around us.

6

The Antarctic Ice Floe

White on Blue

For hours now the *Polarstern* has been encountering more and more ice as it moves southward. The first were pieces floating freely, gently rocked by the swell, then the ice was more and more compact, forming bands of a few hundred meters that alternated with the open sea. Now a continuous white blanket leaves just a few chinks of free water here and there. We are beginning to hear the roar of the icebreaker moving smoothly through this incipient ice, which is gradually becoming thicker as we head toward more southerly waters. People gather in the spacious cockpit to watch the spectacle of birds flying over the margins of the ice floe, searching for food at the interface between the ice and the open sea. The first expeditions to Antarctica were surprised to see in such an apparently hostile environment a large number of whales, penguins, seals and flying birds diving tirelessly near the ice. As early as 1942 it was realized that life depends on the formation and disappearance of the sea ice, on its tireless dynamics of expansion and retraction.

Sea ice forms every year due to the cold in the Southern Ocean. The closer you are to the continent, the more likely it is to find ice that has accumulated over the years, forming thicknesses of up to 3 m and impenetrable even to the mighty *Polarstern*. "Satellite images processed with 3D software have allowed us to calculate a thickness ranging between 1.4 and 5.5 m thick," says Dr. Williams of the University of Tasmania, "but sometimes it can be up to 16 m thick in certain areas of the continent." The area of frozen sea that we could call 'permanent' (and this has wide fluctuations) is about 4–7 million square kilometers, about 14 times the area of Spain, bigger than the whole European

© The Author(s), under exclusive license to Springer Nature Switzerland AG 2022
S. Rossi, *A Journey in Antarctica*, Springer Praxis Books,
https://doi.org/10.1007/978-3-030-89492-4_6

Union. This is in midsummer, around December to January, when it has been melted by continuous solar radiation and milder temperature conditions than in winter. In fact, only 3% of the surface of the sea containing the Antarctic ice floe is completely free of ice.

As the sun begins to disappear, temperatures drop dramatically and the seawater begins to freeze again, covering an area that can easily quadruple the area that remains frozen in summer (it can reach some 30 million square kilometers on occasion, but is usually about 20 million square kilometers, bigger than Russia). It is a process that can occur either abruptly or more slowly, depending on the weather and marine conditions (wind, cold, currents, etc.). It begins with the formation of a thick slush, which is followed by small spangles that then increase in diameter. More or less abruptly it takes the form of large 'cookies' (or 'pancakes'), before coalescing into wide plates and then a more uniform layer. In the end, the ice thus formed will possess a salinity of about 10‰ or less. In certain coastal areas, the formation of ice floes that extend to the bottom of the sea is not an uncommon phenomenon, trapping organisms. They are surprised by a kind of stalactite growth that can reach 30 m down in record years, thanks to changes in salinity and movement from the upper to lower water masses. "The limit of these icy columns is the pressure itself, given by the depth," observes Paul Dayton of the SCRIPS institute in California; "beyond 30–35 m you cannot agglutinate the crystals that allow the formation of these strange columns."

The light gradually becomes unable to penetrate to the floe's lower parts, but nutrients accumulate and are used by the microscopic algae that receive just enough light to continue functioning for a few more weeks. The ice, especially any accumulated snow, blocks the light essential for the photosynthetic machine to function. This restricts their environment to that part of the ice where euphausiids (such as krill), copepods and other herbivores remain, increasingly lethargic, in these lower parts.

For obvious operational reasons, it has always been difficult to study the dynamics of plant and animal communities in autumn and winter, but we know enough to say that there is a profound shutdown governed by this absence of light. At the surface, most of the organisms that live by hunting for fish (such as seals, penguins and whales) migrate toward the sounds of plate growth; that is, northward, out of the vast desert of frozen ice in which barely anything lives. This movement lasts until spring, when the melting ice again gives them the opportunity to move inland, toward the pole. When the ice melts, the penetration of light and the abundance of nutrients causes a strong reaction in the phytoplankton, and its explosion in productivity draws other organisms directly or indirectly to feed, reproduce and grow. It is a short period of time (in certain areas, just a few weeks), but intense. The ice, influenced by the sun that regulates everything, governs all activity in Antarctica. It controls life and death, and represents a force that both accumulates nutrients and phytoplankton and slows down organisms' physiological functions.

Being close to the ice means more opportunity to proliferate. In fact, it is near the ice where we find whales, penguins, killer whales or fulmars, because this is where there is krill, consequently microscopic algae and small copepods. As we see later, it is the extent of the ice that determines krill's presence, extent and abundance. The deeper the ice, the more it is multiannual (more centimeters depth means more years of accumulation), hence the more krill and associated life.

As expected, the circulation of the Southern Ocean is what shapes the extent of the ice, and therefore of life, in this part of the planet. As we move away from the ice edge, the concentration of chlorophyll, nutrients and food decreases dramatically. Ice that is in contact with water features small channels in which chlorophyll can reach values of more than 400 mg/m^3; 30–40 cm below the ice, the concentration is still high but several orders of magnitude less (about 4–5 mg of chlorophyll per cubic meter) than in these labyrinths of life, in which countless organisms thrive.

Understanding the variation in the extent of the Antarctic ice floe will indicate how climate change is influencing the productivity and therefore the dynamics of other organisms, their distribution, survival and displacement. Seals, whales, birds—all await the manna of the ice that fluctuates seasonally and is also dependent on factors sometimes far away from the white continent, as we will see later.

The Mechanism of Life Associated with Ice

Let's take a closer look at how the photosynthetic machinery works in the ice. On the 2003 expedition, we installed an underwater robot just below the ice floe and could observe the numerous curtains of algae hanging from the underside of the ice. Diatoms mostly, so-called algal 'fluff', covered the ice surface irregularly, with krill agglomerating in clouds of various sizes. The camera confirmed that the light was dimmed by the thickness of ice (1–1.5 m), blocking its full intensity. Algae are found beneath the ice, especially within the small features that create about 4 m^2 of surface for each kilogram of ice, forming meandering channels, nooks and labyrinths filled with flora and fauna over more than 40% of its surface. A microcosm is created within which are predators and prey, decomposers and parasites, autotrophs and heterotrophs—a true miniature ecosystem of extreme variability. "If we take two ice cylinders at a distance of 1 m from each other," says Susanne Fietz of Stellenbosch University, "the variations in their chemical compounds can be huge."

The physical and chemical conditions of the environment create, in fact, a series of layers that may have little to do with each other. Small variations in nutrients, in incident light or in certain types of physical conditions such as currents, salinity or temperature will favor one group of algae over another. A species that does best with high nitrates and low phosphate may be inhibited if there is too much dissolved phosphorus or if the salinity does not create the optimal conditions for growth. In these channels, dozens of species lie dormant, waiting for their time to grow. The size range of organisms under the ice can go from smaller than a micron (in the case of viruses and bacteria) to several centimeters (such as krill), and will depend greatly on each area, the type of ice, the time of year, and so on.

Let's get rid of the idea of homogeneity, that everything is the same under the ice, because that is far from what we find in Antarctica. In this underworld, the most active part is undoubtedly the interface between ice and water. And within this interface, the so-called 'fast-ice' is the most prolific. This is the type that forms every year anew without accumulating multi-year layers, making highly dynamic areas. It traps and releases nutrients and algae, creating a low-temperature organic broth in which diatoms and other algae can withstand the temperatures of less than -1.7 °C.

It should be noted that, due to the presence of salts, seawater can drop to temperatures of -1.8 or even -2.0 °C. Due to the moraine incorporated within the ice, salts (highly concentrated) and organic substances allow the interstitial water to remain liquid at yet lower temperatures (about -8 °C is normal), in which algae and other organisms survive until they are released by the thaw. As expected, these organisms have evolved to create cryoprotectants, the molecules that prevent freezing, to ensure the perfect functioning of the cytoplasm (i.e. the aqueous interior of the cell). Diatoms, flagellates, heliozoans and foraminifera are capable of slowing down the formation of crystals within their cell structure by concentrating certain organic substances on the outside, at a rather high cost in terms of energy. An accumulation of lipids in their interior, thanks to its functioning in times when there is an abundance of light and nutrients, allows them to cope with this and other energy costs, such as breathing. With just a tiny amount of light many species are able to continue to photosynthesize, adapting to the lack of photons and compensating for the respiration that every living organism needs to perform.

But, even within the main ice floe, not all parts are covered by ice. The presence of seamounts and katabatic winds (as we have seen before) creates polynyas, a kind of lake whose shores are formed of ice. The wind pushes the ice, and those currents that collide with underwater protuberances create an upwelling of water that can keep an area free of ice for years. One of the best studied and most extensive is the Weddell Sea, about 200,000 km² (almost half that of Spain), associated with the Maud Rise seamount. There, even in winter, seals and whales congregate in search of food. Polynyas are places where the water loses heat and can 'steam' if the outside temperature is very low. This should be no surprise, as the air temperature can drop to -40 °C yet the open sea will never go below -2.0 °C, so is warm in comparison.

Apart from their biological function, these inland lakes in the vast expanse of ice are sources of cold water that then sinks and joins the deep Antarctic water, which will travel thousands of kilometers to emerge later and warm up in the endless cycle of ocean currents. However, something is changing on the Antarctic ice floe. Gradually, the effects of global warming are beginning to be seen, and they are most marked at the poles. Current models predict a decrease in ice cover of between 2.5 and 3 million square kilometers in just a few decades, especially in certain areas. The dynamics of the ice is being altered, and with it the associated life.

Ice Comes and Goes

What happens in Antarctica affects us, and also the other way around. The Antarctic ice floe is influenced by regional phenomena and in addition by more global climatic phenomena. An example of what I am saying is the El Niño and La Niña phenomena, which originate in the waters of the South Pacific and affect the entire planet. To summarize a great deal, El Niño occurs when the coastal winds that carry the waters off the coast of Chile, Peru and Ecuador out to sea weaken, allowing in a large tongue of warm water mainly from Indonesia and Papua New Guinea, transported as a Kelvin wave to reach the coast, replacing the area's cold waters. These Kelvin waves are produced by a change in water temperature that induces an imbalance (something similar to a slope), causing the massive displacement of water from one point to another until it encounters an obstacle (such as the huge current from the western Pacific that crashes into the coasts of Peru and Chile). The propagation and arrival of this Kelvin wave, this immense mass of warm water (up to 8° warmer than the average temperature in these parts), prevent these deep, cold waters from surfacing. The system is no longer supplied, and its production is transformed.

But the El Niño phenomenon has many other effects, including heavy rain in the coastal zone of the Andean Cordillera, extensive fires in the Amazon basin due to drought and a decrease in the frequency of hurricanes, among other climatic anomalies across the planet. In contrast, the phenomenon known as La Niña has the inverse effect, and it is even less understood: there is a drastic drop in water temperatures, which also alters the coastal circulation system and also both the local and global climate.

How does ENSO (the El Niño Southern Oscillation) affect the Antarctic ice floe? Consider that the distance between the American continent and Antarctica is just over a thousand kilometers, so a warm water current arriving from the area of Australia and Indonesia can be partly diverted by South America's Cape Horn to the south. The 'impenetrable' circumpolar current cannot avoid being influenced by the warm water mass. When the phenomenon is strong, the Ross and Bellingshausen Sea area (to the west of the Antarctic Peninsula) loses ice, while the Weddell Sea (to the east of the Peninsula) gains it.

This heat balance is reversed when the phenomenon is La Niña. The colder-than-normal water masses increase the extent of ice on the western side of the Peninsula, while the eastern side is influenced by higher temperatures and loses ice coverage. The influence can last for several seasons. The greatest anomalies are detected in the austral winter, when the extent of ice is greatest. Understanding the flow of warm and cold water (and air) masses is the key to understanding these trends. "El Niño phenomenon affects the wind that influences the Antarctic Oersted ice sheet, which is exposed to the penetration of a warmer water mass," says James Pope of the British Antarctic Survey; "these thermodynamic processes that change the currents are partly responsible for changes in the extent of the ice sheet, not just local effects."

We do not know if ENSO is increasing in frequency and intensity due to climate change—it is still too early to assert this trend. But it is clear that climate is influenced by this phenomenon all over the planet, and Antarctica is no exception.

7

Microbes

The Smallest, the Most Important

When we think of Antarctica, in our mind's eye we always see penguins, seals and whales. The large organisms are the ones that grab our attention, and we tend to despise the minuscule for being microscopic and insignificant. However, as Pep Gasol, a colleague from the Institute of Marine Sciences of the CSIC (Spain), pointed out to me during the 2000 *Polarstern* expedition, microorganisms' biomass (and importance to life cycles) is far superior to that of these warm-blooded animals or any cold-blooded creature. It is just the fact that we do not see them that makes their basic roles, which are so many and varied in this remote part of the planet, imperceptible to us.

Microorganisms dominated the Earth for hundreds of millions of years before the metaphytes and metazoans arrived. Algae and protozoa are as important as bacteria, archaea and viruses in a complex underworld that is hidden from us and that only recently could be scrutinized with the right tools.

At the beginning of the twentieth century the *Discovery* (1925–1937) undertook the first surveys of life in the waters around the white continent. The bacterial component totally evaded this group of enthusiastic scientists who, fighting against all kinds of climatic calamities, were trying to unravel the basis of the functioning of the pelagic and benthic ecosystems of its icy waters. In a superhuman effort they began to collect, separate and classify phytoplankton and other unicellular organisms visible using the microscopic techniques of the day. However, no proper observations could be made of anything smaller than 20–40 μm, a size that their optical microscopes could barely see, to describe the beautiful forms of planktonic algae. Many of them

© The Author(s), under exclusive license to Springer Nature Switzerland AG 2022
S. Rossi, *A Journey in Antarctica*, Springer Praxis Books,
https://doi.org/10.1007/978-3-030-89492-4_7

surely sensed that the same bacteria as in the biogeochemical cycles of the terrestrial, fluvial and marine ecosystems of their origin (in the United States and Europe) were invisibly governing those that they were beginning to study in the Southern Ocean. Considerable yet uncoordinated advances were being made regarding the role of these microscopic organisms, but it was not until the start of the BIOMASS program, in which an international team began to analyze the various components of the water column (presence, diversity, biomass and production), that integrated data were obtained on life cycles in this water. "Satellite imaging results of changes in the ice sheet began to mesh with those of primary production," says Sayed El-Sayed of Texas A&M University, "giving numbers of average variation in ice cover from about 4–20 million square kilometers and the average phytoplankton production itself, about 30–40 g of carbon per square meter per year on average." Although studies on bacteria and so-called picoplankton (eukaryotic cells smaller than 2 μm) had begun, their role was far from being defined.

One of the first steps was to understand the diversity of these tiny inhabitants of the planet, something that began to be a reality in the early 2000s with new genetic techniques. "With the new methodologies, a very high diversity of prokaryotes and eukaryotes never before revealed has been detected," said Carlos Pedrós-Alió of the Institute of Marine Sciences of the CSIC. In the Arctic for the first time, together with other researchers this scientist began to unravel the hidden side of what until then had appeared as an undefined set of organisms in an apparently homogeneous and not very diverse soup. "The diversity even in a place as devoid of nutrients, with as little life as the Polar front in the middle of the Atlantic Ocean, is staggering," says Pedrós-Alió (Fig. 7.1).

The new genetic tools, perfected day by day by several research teams, made it possible to understand that we are not in a soup deficient in species but in an intricate underworld that, due to its diversity, allows a balance and rebalance according to the specific conditions prevailing at each moment. Herein lies the grace of diversity, whether of organisms as primitive as archaea or as recent as primates: environmental changes, continuous both in seasonality and by the effects of climate, give an organism that has lived comfortably up to that moment a hard time if—to give some examples—the amount of nutrients or the light or the acidity of the water change. Then the necessary conditions are created for another organism to emerge from nowhere, cornered as it was, and begin to grow and occupy, in part, the space left by the other. Recent research is showing that this diversity has implications for the functioning of biogeochemical cycles, in which, in an infinite loop, matter changes its form and changes hands among the multiple links of Antarctica's trophic

Fig. 7.1 Rosette ready to be deployed to sample water and physical parameters

networks, and in its icy waters the role of microorganisms is essential to the functioning of its complex natural machinery.

In fact, Antarctica is a powerful focus of attention to understand evolutionary radiation and the creation of so-called endemism (living organisms that, being isolated for a long time, create their own evolutionary branches, disconnected from the outside). Places like Lake Vostok, buried by ice for millions of years, are ideal for trying to understand key aspects of the radiation of eubacteria and archaea (with their problems, as seen in a previous chapter). These areas have been sealed for very long periods of time, so are a perfect place to study evolutionary processes (and times of "creation" of new species). However, more accessible places are also perfect for such studies. "Different environments have the ability to select for different types of microbes, depending on their environmental stability and connection to the outside world," says Warwick Vincent of Laval University, Quebec. The important thing is to try to quantify this potential connection or disconnection with the outside world, and to draw an evolutionary line (or, rather, a sinuous curve) that allows us to know where the different species are located. We know, for example, that the Polar Front was created approximately 10 million years ago, at which time Antarctica's climatic conditions were transformed and its isolation from more northerly currents increased. "That is why it is assumed that in several places in Antarctica endemism must be at a maximum with respect to other more interconnected places on the planet," emphasizes Vincent.

But Antarctica is no longer as isolated as it once was. Apart from long-distance transport by wind (which has always taken place, with a greater or lesser intensity depending on climatic conditions), bacteria and other microorganisms are transported by other means. Mammals and birds make long migrations, so the barrier of the Polar Front is being overridden again and again. The difference is that these creatures may now be carriers of microbes and pathogens that until recently they did not encounter, which have come to them, in their turn, from distant places. We already find fungi such as *Penicillium*, probably transported by man himself or other vectors that apparently were not there before. Due to the climatic rigors, many do not survive, but others do. The exchange of microbes can lead to infections that spread rapidly and uncontrollably between animals or plants. "There is an urgent need to understand the microbial systems on the white continent before they are irreversibly damaged," concludes Vincent.

At the end of it all, in this invisible world, are viruses. Undoubtedly, viruses are also a fundamental part of the functioning of ecosystems here. However, the work done on this remote part of the planet is very limited. What has been found, as elsewhere, is that the abundance of viruses is closely related to that of bacterioplankton. "It is estimated that 10–20% of bacteria are 'lysed' (broken) per day by viruses, i.e. they die by their direct action," writes David Pearce of the British Antarctic Survey, Great Britain. This scientist also points out that 30–50% of new viruses may be produced every hour in this part of the planet, and that viruses infect more than 17% of bacteria. "Viruses can influence the interactions of food relationships, being more lethal to the bacteria themselves than other organisms that eat them," concludes Dolors Vaqué of the Institute of Marine Sciences of the CSIC. Vaqué has been studying the life cycles and the importance of Antarctic viruses (and those of other sites) for a long time, reaching the conclusion that viruses may possibly have a greater regulatory importance in this area of the planet than in other places.

In general, many species of microorganisms found in Antarctica are found nowhere else on the planet. Many species of archaea, for example (bacteria very different from those we are used to hearing about in our day-to-day lives) are specific to these environments, the ice being a stable colonization platform for those accustomed to conditions of the rigorously low temperatures and the intermittent lack of light. What seems important is that, to a greater or lesser extent, these microorganisms have a fundamental role in the decomposition, recycling and transformation of matter, even though they can evade detection, and even more so regarding their role in the system.

Water Warming and Microbial Biota

We are beginning to understand that climate itself depends on the biogeo-chemical cycles of the Southern Ocean. What happens there at the level of microorganisms (from primary production to the recycling of part of the matter by bacteria) has repercussions on the rest of the planet because of their role (photosynthesizing or not) in these remote and cold waters. So, what is implied by a change in temperature, currents or pH of the waters that host them? How would this community respond to climate change? "We know that general models predict major changes in ocean circulation in this part of the planet," says Rainer Zahn, a specialist in paleoclimatology. These changes will have a great influence on the carbon balances in the area, because as air and water temperatures increase (which is already happening in certain areas), water stratification will increase, reducing the flow of carbon toward deeper areas. Therefore, the capacity to capture CO_2 will be limited, which would only worsen the situation by reducing an important carbon sink.

In short, by increasing stratification water mixing would decrease, putting the brakes on the capture of CO_2, which is already in excess. Positive feedback would be created; that is, resulting in more and more CO_2 in the atmosphere. "Current models estimate a temperature increase of up to 2 °C–3 °C in areas such as the Antarctic Peninsula, one of the most productive areas on the planet," predicts Uli Bathmann of Germany's Alfred Wegener Institute. "Local factors are very important in these cases," remarks Hyewon Kim of Columbia University; "in long time series of 20–30 years we see that in some areas chlorophyll increases while in others it seems to decrease; this is due to the fact that, apart from water warming or pH decrease, factors such as stratification, wind or glacial or ice inputs can locally shape environmental conditions."

The rise in temperature would produce a paradox, as photosynthetic mechanisms will accelerate, increasing primary production (algae proliferation) that captures the CO_2 that worries us so much when it is in excess. "We will have to understand the puzzle well, because the answers to the problem are far from linear," emphasizes Batman.

The dynamics of the ice floe itself, in frank regression in the northernmost (and most productive) areas of Antarctica, will play a role in this complex future scenario. The regression (discussed in the previous chapter) implies more light, but since the cycles of many algal species are linked to the ice itself it is not clear what the final trend will be. For example, nutrients depend on that ice, and changes are already being observed in the cycling of iron, a micronutrient essential for algal blooms. "Phytoplankton depend on iron, but its solubility and concentration are changing rapidly due to acidification, stratification, increased water temperature, and deoxygenation," says Dr. Hutchins of the University of Southern California.

There are further effects. An estimated 10% of the planet's dimethyl sulfides (DMS), partly responsible for cloud production and precipitation, comes from the Southern Ocean. What will happen if the balance of these chemical compounds the changes? "A species like *Phaeocystis antarctica* is responsible for most of these compounds in this area," comments Sayed El-Sayed, "and it is closely linked to the ice dynamics itself." The accelerated changes will lead to transformations that are not possible to predicted well, at the moment. Finally, the ozone depletion in this part of the planet also affects algae and other microorganisms (and of course macroorganisms), and the time window in which the greatest depletion occurs is in September to October, just when the photosynthetic machinery is being reset (due to the enormous extension of ice over the sea at this time in this part of the planet).

What effect will an increase in temperature and stratification have on bacteria? It has been found that most of the species studied are accustomed to cold conditions for the optimal use of the organic matter that they have to decompose. An increase in temperature could greatly accelerate this decomposition, transforming part of the organic matter into CO_2. "This could be especially acute in deep areas such as the seabed, where large amounts of intact matter accumulate," explains Enrique Isla of the Institute of Marine Sciences of the CSIC. At 150–300 m, a large number of organisms depend for primary production on the rain that sinks periodically from the surface. As we will see later, these organisms, very rich and diverse, rely on these pulses of new food creation, which flow both up and down and can be transported laterally by currents. If temperature were to increase and with it the efficiency of bacterial respiration, decomposition would accelerate and less matter would be available for the sponges, gorgonians and a long et cetera of bottom dwellers. Besides that, there could be situations of hypoxia, where matter accumulates more because there is a greater rain of microalgae, thus many organisms could suffer from an oxygen deficit and then disappear.

Changes in ocean productivity are already occurring, and Antarctic waters are no exception. "We need a good understanding of the 'sequence' of factors that are going to influence that productivity, such as warming, pH, ice, etc.," explains Stacy Deppeler of the University of Tasmania. "The Southern Ocean is a very vast and complex place, we tend to lump everything together and that's a huge mistake." All the research teams are deeply concerned with the sequence of events that are already occurring. In a recent article, several scientists from all over the planet agree: "We need to understand how microorganisms affect and are affected by climate change." The results are not yet clear, and the responses of these microorganisms are heterogeneous. It is a complicated picture, full of factors that are linked to each other, like everything else in this and other ecosystems.

The LOHAFEX Experiment

In the face of rising temperatures and greater acidity in the oceans, several projects have been launched in which so-called 'environmental engineering for climate change mitigation' is applied. These large-scale experiments are not without controversy, and the latest LOHAFEX project, undertaken from the German oceanographic vessel *Polarstern*, was in the polar waters of the southern hemisphere. Professor Victor Smetacek of Germany's Alfred Wegener Institute, in collaboration with an Indian research group led by Professor Wajih Naqvi of the National Institute of Oceanography in Goa (India), is one of the main promoters of this idea. Algae, especially diatoms, so important to marine production, have elements that limit their growth. Among these are micronutrients.

Micronutrients, unlike nitrogen or phosphorus, are really very scarce and are present in very low concentrations in the sea, yet are essential to cellular functioning. One of them is iron (Fe), whose importance to the production of these algae was demonstrated some time ago, as mentioned earlier.

In the experiment carried out between February and March 2004, soluble iron was added to create a major proliferation of algae that would be consumed by small planktonic crustaceans or, after having absorbed CO_2 during growth, fall to the sediment below. According to the promoters of this large-scale experiment, its aim is the sequestration of CO_2 by microscopic algae to help mitigate climate change by locking up some of the carbon in deep seas. Early results indeed showed a major algal bloom and an increase in the concentration of copepods and other small crustaceans. "A very large diatom bloom was created in five weeks," Smetacek says, "of which 50% of the carbon could have been exported to the bottom." That is the purpose, to capture atmospheric CO_2 and immobilize it for hundreds of years at the bottom, at a depth of more than a thousand meters.

But there are voices within the scientific community that strongly oppose this environmental engineering tactic. At a congress held in Nice in January (ASLO 2009) Professor John Cullen of Dalhousie University in the United States pointed out the fact that little is known about the effects on the rest of the ecosystem of this type of experiment: "It is quite possible that what falls by gravity returns to the system breathed by bacteria (which are the main producers of CO_2 on the planet), so we would not be doing anything." Several experts have conducted fertilization experiments (in the Pacific, in the North Sea and off the coast of Ireland), and have found something unpromising: by supplying iron, a toxic species of diatom seems to prevail over the others (*Pseudo-nitschia variabilis*), forming extensive 'red tides' whose effects on the ecosystem have proven harmful. "We should not be tempted to experiment without absolute certainty that we will not harm the environment with our 'remedy'," commented Ken Caldeira of Stanford University's Department of Global Ecology, at the same ASLO session: "We would only need the public to see marine biologists as 'the bad guys'." Smetacek and Naqvi defend themselves, saying that they understand that there are associated problems but that we should give this type of geoengineering (at the planetary level) a chance, because we know very little yet: "All the critics focus on the worst-case scenario." Naqvi stresses that "more experiments are needed to understand whether or not it is feasible and effective." Both Smetacek and Naqvi conclude that experiments must always follow protocol and be controlled by specialists.

Perhaps we should approach it a little differently. Descending 1000 or 2000 m down to the bottom involves a great deal of respiration by organisms in the water column, including bacteria, protozoa and of course copepods. Some organisms at the bottom of the sea immobilize carbon for centuries or millennia, and these might help: the benthic suspension feeders (corals, sponges, gorgonians, bryozoans). These organisms feed on microalgae, detritus and zooplankton. If the bloom were to occur in shallower areas (such as the continental shelf) and all these "mouths" were waiting, the biomass would arrive faster. Who knows… maybe we could have a very efficient 'blue carbon' generating machine that would help us to mitigate the effects of climate change a little. The carbon would stay in these organisms' structures and surroundings, promoting the sequestration of part of the excess carbon produced by humans.

8

Ice, Algae, Krill

A World That Revolves Around Algae and Zooplankton

Antarctica is undoubtedly one of the places in the world with the highest production of microscopic algae. Concentrations can reach more than 8 mg of chlorophyll per cubic meter in surface waters, thanks to the blooms, or population explosions, of diatoms or *Phaeocystis antarctica* (the two groups that dominate the microalgae scene in this area of the planet) that recur in several areas of the Southern Ocean. Over a few weeks, actual fronts of more than 100,000 km^2 in size are created, which can be photographed by satellite, feeding a huge amount of biomass in the form of protozoa, copepods, krill, fish, penguins and seals. It is important to understand that this area of the planet absorbs a large part of the planet's natural CO_2 and of anthropogenic origin, ending up in the many links of the food chain or immobilized on the seabed as sediment. In fact, in the area of the Peninsula, production is so high that sediment at a depth of 1000 m in the Orleans Canyon smells rotten due to the sheer volume of decomposing algae that has descended through the water column to the bottom.

In all this equation, ice plays a major part, both in its formation and in its melting in seasonal cycles that control the survival of species. As we have seen, it releases algae and the upwelling system produced by movements of the water masses provides nutrients in abundance. Despite the low temperatures, algal growth is favored by the plentiful light and nutrients. In this part of the world, because there are no large rivers to transport it continuously from land,

© The Author(s), under exclusive license to Springer Nature Switzerland AG 2022

S. Rossi, *A Journey in Antarctica*, Springer Praxis Books, https://doi.org/10.1007/978-3-030-89492-4_8

one of the deficiencies is iron. "Iron limitation," says Viktor Smetacek of the Alfred Wegener Institute, "is fundamental to understanding how the system works; the planet's phytoplankton is responsible for the transformation of about 50×10^{15} g of the planet's CO_2 per year, and in this cold southern zone one of the factors limiting the already exacerbated growth of phytoplankton is the lack of this microelement."

The size of the algae also matters. This is also due to the dynamics of the ice, depending on the extent of both the ice coverage and the stratification of the water. In more stratified conditions diatoms dominate, which can be large or small, with more or less silica in their external structures (the frustules). Having more silica, being larger, means more work for the tiny copepods that eat them, which find them unmanageable and indigestible. If the water is less stratified and more mixed, blooms of different algae are created, such as small diatoms and the *Phaeocystis* we were talking about before. The copepods (which can constitute a biomass of more than 50% of the zooplankton) are happier in these conditions because they can eat the algae more efficiently. These small crustaceans (*SpongeBob*'s famous Plankton character is a cyclopean copepod, saddled with a bad guy role) break up their prey with their jaws before ingesting it.

While copepods are selective about their prey, the other essential component of crustacean plankton, krill, are less fussy. They are small shrimp, about 4–8 cm long, that eat a little bit of everything (even their own larvae, if they find them), so they make fewer distinctions when processing material. Krill catch prey ranging in size from six microns to a few millimeters, browsing here and there on what's within reach. "Krill transform and transport essential nutrients and leave them dissolved throughout the water column," says Dr. Covan of the University of Tasmania. "Their role in the biogeochemical cycles of the white continent is essential." Algae and zooplankton are at the base of the "chain of the giants", because they are the ones that feed the whales, seals and penguins.

Along with colleagues, I have tried to understand how this complex machinery works. We have been able to see that there are about nine or ten annual production pulses in the Weddell Sea, for example, that favor the growth of algae. If not consumed at the surface, the algae are transported to the bottom of the sea to feed the suspension feeders, detritivores and any other creature prowling around the area, whether directly or indirectly. But before the algae arrive, they have to pass through a filter, a 'customs point' constituted of copepods and krill. These animal plankton feed, reducing those eight milligrams of chlorophyll per cubic meter to less than four—yet they can't manage them all. The more algae there are, the more is eaten and what is defecated remans almost intact and thus is of less use to the herbivores responsible for cleaning the area. The food reaches the bottom directly, by sedimentation of uneaten algae, or as organisms' feces.

Everything is good for those below, waiting for the manna from the ice. Recall that some of the krill 'poop' is recycled at the surface, and such a surplus indicates that everyone is happy. "If you compare blooms of diatoms and *Phaeocystis*," says Kevin Arrigo of NASA's Goddard Space Flight Center, "the latter type of algae exports more carbon to the bottom." This point is not at all superfluous. If the water becomes more stratified due to global warming, the amount of carbon that sinks to the bottom will be less, to the detriment of the species that live below and sequester it in their structures, and less will remain sedimented for thousands or millions of years. The mucus outer cover of these algae helps them to fall about 200 m/day to the bottom. Knowing that the dynamics of CO_2 depends largely on the Southern Ocean, we must keep an eye on who dominates in the near future, due to the changes that the area is undergoing and how these may affect not only the fauna of the area but the rest of the planet.

The Fundamental Role of krill

Stocks of krill feeding on algae in the Antarctic, according to estimates, range between 80 and 1 billion tonnes (in an area of 8.4–12.5 million square kilometers)—too wide a range, and imprecise. You can read, read, read and read to see what everyone says about it. Why the hell are we not able to calculate precisely how much krill there is in this area of the world? Krill have been a nightmare for many scientists who have tried to predict their numbers, movements and ecology. The topic has been studied since 1920, and in 1981 the BIOMASS program was launched, an ambitious study that sought to understand krill biology, ecology and stocks, in order to catch it. But no matter how hard they tried, scientists have been unable to find the key. What factors did the project find relating to higher or lower concentrations of krill?

This is an interesting point, because it involves the numbers (as we will see later) able to be caught while respecting the actors in the ecosystem that depend on it. Let's remember, for example, that more than 90 million tonnes of fish are extracted from the sea every year (extractive fishing only). It is tempting to think that, with our current collapse of fishing, if there are really 1000 million tonnes (which seems to me a bit exaggerated) we can rely it as a protein source. However, it is clear that there is plenty of krill, because what whales alone consume far exceeds the imaginable. A conservative estimate argues that when whales were abundant in this area of the planet, they needed approximately 190 million tonnes of krill each year to feed on, which would mean a total stock of about 600 million tonnes to be able to be a viable crustacean population. Now that there are fewer whales, stocks should be greater, but things are changing (as we will see), and krill seem to be decreasing. In addition, other predators such as penguins or seals have increased their numbers, taking advantage of the absence of whales, interacting with the small shrimp yet partially (never totally) replacing the cetaceans. They interact, but they do not control them. It seems that it is not predation that is limiting this organism's stocks but its capacity to recruit new individuals, and this depends partly on the ice.

Here comes the complexity of the matter. Attempts have been made to relate krill abundance to temperature, bathymetry, salinity, chlorophyll concentrations, laying sites of new recruits... but nothing. There is nothing conclusive or clear. But there are some clues that let us understand where the shots are going. We know that after a long winter, when the ice melts, the krill 'wake up' and move toward the food source (the surface). "Krill larvae live under the ice," observes Dr. Melbourne-Thomas of the Australian Antarctic Division, "a critical moment in their life cycle." They feed and reproduce, and then in autumn they go back down deeper. They do not stay in the same place where the larvae are. There is a spatial segregation, thanks to which there is not too much cannibalism: they take no prisoners in such a harsh environment, and your offspring's as good as anything else for a creature as voracious as krill. Come winter, there is a lull in activity. "In autumn and winter, when the ice covers much of the Southern Ocean again, krill doze, breathing 30–50% less than in summer conditions, reducing their motor activity by up to 86% and consuming their reserves, especially those they have accumulated in the form of lipids," says Bettina Meyer of the Alfred Wegener Institute. It is assumed that less ice would have a negative influence on the number of larvae but, according to Dr. Melbourne-Thomas, this may not be so: "the higher water temperature may favor the survival of larvae," he says, "and in a few decades, for example, on the west coast of Antarctica it could rise by about two degrees Celsius."

And the cycle starts over and over again, but in different ways because the ice conditions are also different. What the specialists do seem to be clear about is that the more ice there is, the more successful is the recruitment. In other words, if winter comes to cover those almost 20 million square kilometers, there will be more individuals freshly baked, ready to eat algae. Why? On the one hand there is the fact that more ice means more algae 'seeds' for the following year; and on the other it has been observed that the larvae find more shelter from predation and food to get them through the hard season. If there is little ice, a different, more opportunistic organism dominates the scene: salps (chains of gelatinous barrels capable of filtering algae at high speed). It is not that the krill disappear but that these transparent urochordates take over the food, compete with the krill, and thus displace them.

It is a difficult balance, and in the most productive areas (south of the Atlantic, where 58–71% of the production is concentrated) the ice is in regression year after year. Already in the 1970s the decreasing number of krill was detected, but the extreme variability made it difficult to estimate reliably. The trend continued, and although ice cover was linked to the El Niño phenomenon (the 7–10 year cycles are observed also in krill), this was not enough to explain the rapid decline in stocks.

And this decline will affect the rest of the food chain. Even small area shifts driven by deviations in currents or ice cover will cause the abundance of krill at a regional level to change and become less accessible. In the Peninsula, the temperature increase in the atmosphere since the 1950s has been about 5–6 °C. That's a lot. It is already affecting the whole community, in one way or another. It is possible that the krill will migrate to where the ice is; that is, to the south. But there the light and productivity are different, and it is not known what this implies for the whole system as a whole. A lower primary production due to stratification, a smaller krill presence and a less dynamic system imply profound changes in the biogeochemical cycles of the area. Positive feedback might be created, causing CO_2 to increase in the atmosphere as it is retained less in the oceans by biota (thus creating different balances).

And there are further factors to consider, in this story. Krill often spend long periods near the coast, where glaciers are melting faster and faster. These glaciers, as we discussed earlier, carry with them a large amount of inorganic, lithogenic material. "It has been proven that the large amount of 'mud' that the glaciers carry to the sea clouds the water and causes the krill to collapse and die, becoming stranded on the beaches," explains Veronica Fuentes of the Institute of Marine Sciences of the CSIC (Spain); "from 2003 to 2012 it has been increasingly frequent to find dead krill on the beaches of the islands near the Peninsula, where practically all the glaciers are in rapid regression."

Temperature, productivity and ice are the key to understanding krill dynamics and all the consequences on Antarctic food webs. "There will be areas where it is more abundant, others less, but any fishing model will have to be compatible and flexible with these complex dynamics," says Andrea Piñones of Yale University. All agree on the fact that the drop in krill production may lead to difficulties for whale or penguin populations, for example.

"Our models indicate that the long-term decline in krill will have a strong influence on some penguin species," explains Lucas Krüger of the Chilean Antarctic Institute; "less ice, less production means that penguins can feed their chicks less well, which implies a gradual decline in their populations." And not only that. In these models, whales are affected when climate change and krill fishing in Antarctic waters coincide. "After being nearly extinct, we thought that Antarctic whales now had a chance as there was more krill for fewer individuals," says Vivitskaia Tullok of the University of Queensland, "but fishing and the increase in their competitors for food, coupled with climate change, jeopardize this prospect in future decades."

Let's remember that krill is not just a 'shrimp', not just a small crustacean. It has implications for the regulation of biogeochemical cycles and carbon sequestration. Krill, *Euphasia superba*, is a crucial part of all of us, even if it doesn't appear that way. In short, an apparently insignificant organism plays a vital role in Antarctica and the Southern Ocean according to its ability or inability to help to regulate the planet's CO_2, in part defining our destiny.

9

Life in the Dark: A Diversity Explosion Under the Ice

Surviving in the Depths of Antarctica on Ice Manna

We have already seen that in the water column in the Antarctic there are two well-defined seasons of production. This strong temporal asymmetry could be summarized as one highly productive period coinciding with spring and summer, and another when the light begins to disappear, putting to sleep the biological machinery set in motion by phytoplankton, coinciding with autumn and winter. Like virtually every aquatic system, the Antarctic bottom also depends on these algae to function.

Many organisms in winter remain in a state that could be described as lethargic, while others migrate northward to escape the lack of sufficient primary production to maintain their most basic metabolic needs. But in the benthos (bottom) some cannot move, or move little, so they cannot avoid this lack of food. The few yet convincing studies have shown that in certain cases many remain active throughout the year. These are the benthic suspensivores.

As the term suggests, a suspensivore is an organism that lives on suspended particles. It eats them, capturing them in various ways depending on whether it is passive (like gorgonians or hydrozoans, which wait to intercept the prey) or active (like bivalves or ascidians, which pump water into themselves). The point is that, during spring and summer, the flow of live or dead particles from the algal blooms reaches the benthic organisms as a rain that is rich in nutritive substances. It has been found that most of the material caught in scientists' sediment traps (cylinders placed on the bottom to catch particles

© The Author(s), under exclusive license to Springer Nature Switzerland AG 2022
S. Rossi, *A Journey in Antarctica*, Springer Praxis Books,
https://doi.org/10.1007/978-3-030-89492-4_9

falling from the surface) are diatoms, and other phytoplanktonic algae are secondary, such as dinoflagellates. Using devices that measure current and resuspension, it is possible to calculate how much of this material is from vertical transport (from primary production) and from lateral transport (also from primary production but not directly from that taking place above at that moment on the surface, but from other areas, sometimes far away).

Receding ice is the main promoter of blooms, increasing the chlorophyll concentration as much as tenfold just a few days after melting. But, as we suggested earlier, by no means everything stays in the surface column. In fact, despite frantically taking advantage of this algal bloom, many organisms such as krill, copepods or chains of salps are unable to eat all that is produced. The leftovers, together with the feces of the column filter feeders (at the peak of production, these are rich in undigested lipids), form part of the food for organisms living in rich communities at 100–500 m depth. It is therefore clear that, during the good times, the organisms at the bottom of the Antarctic seas also receive a large amount of food.

But what happens when the food supply is turned off? The period of scarcity is long and, until now, it was thought that the only way for benthic organisms to cope was to remain dormant. Andrew Clarke and David Barnes of the British Antarctic Survey have published a paper showing that, despite slowing down, many of the bottom-dwelling organisms in the Weddell Sea do not totally stop eating in autumn and winter. How is this possible? Two reasons may shed light on the secret. First, the huge lipid-rich surplus that remains unused by pelagic organisms near the surface is preserved for a long time due to the low water temperature. It has been shown that in this system proteins dominate, at almost 50%, and second in quantity come lipids, at almost 30–40%. Unlike in tropical or temperate systems, it is not carbohydrates in that second place, as they account for just 20–30%. Therefore, we have very rich material near the bottom, and it has been found to remain little altered throughout the autumn. In other systems, such as temperate and tropical systems, the macroalgae and marine phanerogams provide a large amount of detritus rich in refractory material (carbohydrate compounds and humic substances), material that is indigestible to benthic organisms. But in Antarctica (apart from in the extreme north of the Peninsula and some other specific coastal areas) these macroalgae are not present, therefore the only source of material from primary production is microscopic algae, which also adopt the strategy of accumulating lipids to face the winter season with a guarantee of survival. Proof of this is the high concentration of krill and salps found just a few centimeters above the benthos, creatures that come down to feed on this rich matter near the bottom.

"There is a good relationship between currents, food and abundance of organisms on the Antarctic seabed," says Jan Jensen of the University of Tasmania. Although this relationship is not entirely linear, as Julian Gutt of the Alfred Wegener Institute points out (especially at a scale of less than 2 km), the truth is that food is always a key factor in finding one or another organism on the seafloor.

There is a second reason for this presence of food during lean times: the continuous pulsed resuspension of material near the bottom. Strong lateral currents near the seafloor are very common, and in the areas studied it has been shown that they are capable of maintaining a large amount of available material 50–100 cm above the bottom. We should add that the communities of organisms throughout this area form a complex three-dimensional structure, and that this can retain particles in the environment far more efficiently than a flat substrate with barely any organisms. "Sponges dominate in biomass and structure this area of the planet," explains Emily Mitchell of the University of Cambridge. "The complexity of the habitat is largely due to their presence; if we were to remove them, up to 42% of the other taxa such as crustaceans, echinoderms or cnidarians might decrease in abundance or even not be present." There is greater complexity, greater particle-holding capacity and greater particle capture—and it is a fact that organic matter is present almost all year round, thanks to resuspension. It is therefore understandable that sponges, gorgonians or ascidians do not enter a completely lethargic state during autumn and especially winter, because the matter accumulated during the spring and summer is rich, circulated in frequent pulses and remains trapped in a kind of three-dimensional web for a long time without falling completely to the bottom.

The Surprise of Antarctic Biodiversity

It is nighttime. All the activity on the deck of the *Polarstern* is focused on a single device, the ROV (Remotely Operated Vehicle). This is a robot controlled from a cabin at the stern of the ship where the pilot (either Julian Gutt or Werner Dimmler, depending on the event) operates the dive, taking it to the bottom to film at two, three or more hundreds of meters, as far as its 'umbilical' (the cable that connects the device to the surface) allows. They are looking for fauna that extend hundreds of kilometers over the deepest continental shelf on the planet.

What I see illuminated by the powerful spotlights is of indescribable beauty. The pilot drives smoothly, so as not to disturb the forest of creatures passing before us. There are sponges, gorgonians, bryozoans with strange shapes, with vivid colors and of various sizes. Above are a large number of sea lilies (crinoids) that, upon detecting the apparatus, dance like graceful ghosts (Fig. 9.1). Next to them, a crowd of holothurians extend their mouth parts like flowers to capture the particles that move above the sessile animals in the currents, as these particles are their source of food. Fish, small crustaceans, starfish … I am surprised by the immense wealth that I see. I did not expect it. My idea was something monotonous, a community composed of one or two dominant organisms and the occasional swimmer or crawler prowling around the Antarctic sediment. But this is not the case. I find myself in front of an ecosystem that has nothing to envy a tropical coral reef or the varied underwater landscape of the Mediterranean. From the start, I realize that any scientific research focused on this group of organisms will be worthwhile. And, among other things, I am committed to understanding how they work. Very promising, very encouraging.

On a continental shelf deeper than usual (between 200 and 600 m) due to the pressure from glaciers as these immense masses of ice come and go (especially during the last million years), a multitude of species have lived almost isolated for a long time. More than 10,000 marine species have been counted, and it is conservatively estimated that there may be more than 16,000. On each trip, taxonomists gather material to describe new findings. "In the

Fig. 9.1 Sponge grounds in the Antarctic benthos (Photo credits, Julian Gutt, AWI)

deepest part of the Weddell Sea, the one comprising a range between 800 and 6000 m depth, in a single campaign, of the 674 different isopods found, 585 were new species to science," says Angelica Brandt at the Alfred Wegener Institute. These are the deepest and most unknown type of fauna, but there is no need to go down to the abyssal zones; in each expedition even at a few hundred meters tens, hundreds of new species are recorded, because this is one of the least explored areas of the planet.

It is a taxonomist's paradise. At the current rate of research and collection of material, it will take hundreds of years to describe all the new finds. There is a lack of taxonomists, an undervalued yet essential branch of science dedicated to collecting, ordering, describing and categorizing the animals, plants and other living beings of the planet. Scientists, due to the stupid maelstrom of rushing, publishing and contract maintenance, can't operate any faster. They need time to be able to do a meticulous, precise and laborious job, and one that we are beginning to understand is the essence of understanding the functioning of the planet's ecosystems.

Another challenge facing these scientists in Antarctica is inaccessibility, being off-route from many places. "There are areas that have hardly been sampled at all," Chester Sands of the British Antarctic Survey told me, "such as the interior of the Weddell Sea or the area between the Eastern Peninsula and the Ross Sea." Within the different categories, within the different groups, there are also strong asymmetries in the numbers of descriptions. For example, the cnidarians (gorgonians, corals, anemones) are fairly well described because there has been a school (under Pablo López González of the University of Seville) dedicated to collecting, describing and classifying them; but the ascidians (chordates that live on the seabed, and considered our closest invertebrate relatives) do not seem to have interested many specialists and are more neglected, from a taxonomic point of view.

But what has caused there to be so much diversity in a place like this? As we have mentioned, Antarctica is considerably isolated under the current regime. But it is also a cold yet stable environment—very stable. Even with its strong seasonal fluctuations and the continuous disturbance by icebergs, its temperature and food are predictable. Diversity occurs when there is a predictable environment, when there is stability in a number of parameters and when these are not extreme. The fact that water temperature is very low is not a problem for many invertebrates, which live according to these physical conditions, and what may weigh more heavily in the balance is whether there is plenty of oxygen (which there is) and a regular source of food.

We know little about the interactions between species in this area of the planet, but we understand that, over millions of years, each and every organism has sought a niche, its specialty. As expected, there is no homogeneous pattern of distribution of organisms and diversity. "On a small scale (less than 40 km), this diversity will depend on the productivity of the area," explains Simon Thrush of the National Institute of Water and Atmosphere Research in New Zealand, "whether or not there is a recurrence in the food available; however, if we compare on a large scale (around 300 km), we will have to take into account other factors such as biophysical habitat." These are currents, the lateral and vertical transport of food, the type of substrate, the presence or absence of seamounts, and so on. It is therefore an ideal laboratory for understanding many of the natural phenomena related to the distribution and diversity of species.

The enormous spatial and physical heterogeneity (currents, tides, etc.) makes it difficult to create a unique pattern, as Julian Gutt explains. A few years ago, an unusual concentration of brittle stars and brittle stars was found in an underwater mountain range extending some 1400 km from southern New Zealand to the Antarctic continent (the Macquarie Ridge). They were beneath a very strong current, an impulse of water equivalent to between 110 and 150 times the sum of all the fluvial flows of the planet. It is one of the few places where the circumpolar current is diverted, deflected by this underwater bulge that acts as a barrier from about 850 m deep to 90 m in the shallowest areas. Once again, the richness of the world's coldest waters was demonstrated, thanks to the production and stability conditions, also showing the enormous spatial heterogeneity that dominates the distribution of organisms.

The heterogeneity is great, and this explains in part the existence of several community types. Suspension feeders dominate this area of the planet, due to the stability of the soft substrate, which does not suffocate them as in other parts of the planet; living on a soft bottom is generally a problem for this type of organism. They collapse easily, and the presence of rivers or continental inputs as strong as those found in the rest of the planet means that in other places only a few species are able to survive.

Not in Antarctica; here they are able to form complex structures in which life has diversified. Because, if the sessile fauna is diverse, the mobile fauna is far more varied. In these animal forests of sponges, gorgonians or bryozoans lurk more than 1800 different species of crustaceans, occupying all trophic levels within their morphological possibilities. All of them (mostly amphipods and isopods) have found shelter and food, have been able to specialize and, in addition, do not have large predators. On the Antarctic continental shelf, this absence of large predators (large decapod crabs, sharks, large bony fishes) arose after the Eocene extinction, and they did not reappear. Evidence of their absence includes the abundance of large amphipods and crinoids dancing everywhere. There is also a lack of large herbivores, simply because, since the bottom is very deep in most of the continent and no light can reach it, there are no large algae or underwater plants.

Nonetheless, the process of speciation intrigues scientists. "In order to explain such a wide level of speciation, not only time is needed, but also certain biological characteristics," says Josep-María Gili of the Institute of Marine Sciences of the CSIC; "one of the factors is reproduction; in temperate or tropical zones it is much more accelerated, in Antarctica it is slower, as is its growth. This makes it difficult to understand a radiative theory as it is proposed in other places."

Suspension feeders (gorgonians, sponges, etc.) are highly diverse, even though their food source is similar. They make up the so-called marine animal forest, in which particles feed these organisms, which wait like trees or shrubs, with specific adaptations, such as corals, gorgonians or sponges. Possibly because they have adopted different reproductive and energy storage strategies over hundreds of thousands, millions of years, they are incredibly diverse. Coexistence in a community may be due to the differentiation of ecological niches, but also to the mechanisms of competition between organisms and a certain degree of chance. This chance is provided by icebergs and their massive destruction, which is then used by organisms to start all over again. And no two communities will ever be the same after a disturbance, so more and more complex networks are created. "Since there is no clear hierarchy between species," says Julian Gutt of the Alfred Wegener Institute in Germany, "the difference in characters, however small, is the key to understanding diversity in communities". Much remains to be understood about the functioning of these platform communities, perhaps the last bastion of a truly untouched, large-scale habitat on our planet.

We still have a long way to go to understand the functioning of these communities. In the deepest areas, at more than a 1000 m, there is scarcely any data. "What is clear," says Francyne Elias-Piera of the Universidade de Sao Paulo in Brazil, "is that we need to understand what they eat and in what quantities to enter the complex world that awaits them due to climate change." Because things get complicated, as for other communities.

As we have discussed before, we have for example an increasing input from glaciers of terrigenous material. "It is paradoxical," comments Gastón Urralde of the National University of Córdoba in Argentina; "if there is less ice, phytoplankton productivity increases and the space that can be colonized can also increase, but the melting ice causes an enormous increase in the amount of refractory sediment." In many cases, this prevents organisms that cannot support so much input from the land from surviving. "Despite the limited data we have," says Santiago Pineda-Metz of the Alfred Wegener Institute, "we have detected a decrease in the biomass of benthic organisms in certain areas of the Weddell Sea; suspensivores in many areas have given way to depositivores (organisms that directly consume the sediment deposited on the bottom, not in suspension), it could have lost up to two-thirds if we consider the data of the time series we have from 1971 to 2014."

There is still much to understand: the wider influence of icebergs, changes in the Antarctic ice floe, collapse of glaciers, transformation of primary productivity ... many factors that are influencing the ability of these communities to capture 'blue carbon'. Because, as we have said before, these are our allies in climate change, and their capacity to capture and immobilize carbon is changing very rapidly.

Ice Fish

In the bowels of the *Polarstern*, below the aft deck, Katia Mintenbeck of the Alfred Wegener Institute measures, weighs and samples the strange ice fish that live on the Antarctic seabed. She is one of the researchers trying to understand how rapid temperature changes in the Southern Ocean may influence one of the most endemic fauna on the planet. With the din of the engines in the background, Katia explains to me that the fish of this part of the planet vary little: "only 322 species of the 25,000–28,000 that are known in the whole planet," she emphasizes. Perhaps 30 or 60 species remain to be discovered, but not many more. This is a very low figure if we take into account that the Southern Ocean occupies 10% of the surface of the planet's marine waters. But then she adds that this should not surprise us: those that have remained here possess the physiological characteristics of no other fish in the world. And that causes an important factor that arises in many taxonomic groups in this part of the planet: endemism.

The endemism in the Antarctic fish fauna is throughout 90% of the species; that is, nine out of ten species found here are not found anywhere else on the planet. This is the highest figure for the so-called closed seas, those that for one reason or another are isolated from the great oceans and have little connection with the outside world. Fish are not the only group with high levels of endemism; in sponges 51% of species not found anywhere else on the planet, and other groups such as polychaetes, amphipods, isopods or mollusks have figures of around 57–88% of endemic species, and the strange sea spiders or pycnogonids are in the group with the most endemism on the Antarctic shelf and coasts, at 91%.

Among all the fish characterized by their adaptation to such a cold environment, the notothenioids are the ones that have transformed the most during the different stages of isolation of Antarctica, so Katia tells me. Also called icefish, they are peculiar no matter how you look at them. "In a very cold environment where the temperature fluctuates very little, ice fish have developed a series of incredible adaptations that we don't find in any other group of fish anywhere else on the planet," explains Bo-Mi Kim of the Korea Polar Research Institute. They have no erythrocytes in their blood; the oxygen and carbon dioxide transport is relegated to a complex physiological system. To begin with, their metabolism is very low, very slow, and they have highly developed gills through which oxygen, which is highly concentrated in the waters of this part of the planet, enters a system of capillaries larger than those in other fish, driven by a large heart that pumps harder than is usual. They also breathe through their skin, which is a highly vascularized accessory breathing organ, something that no other fish species on the planet does. Oxygen dissolves more easily at low temperatures, so erythrocytes are a waste of energy. This process is also necessary to host the large molecules, the glycoproteins, which prevent them from freezing in water below 0 °C. These are antifreeze agents, present in the water and in all species of icefish that need an unimpeded flow through their circulation system. Moreover, as most of them are benthic and are trying to save energy, they possess light bones, very few scales and lack one organ typical of other fish groups: a swim bladder. This fish-regulated pouch allows the individual to go to the bottom or to the surface at will, and notothenioids do not need one because they cruise around the bottom.

These icefish have a slower metabolism yet a greater capacity to accumulate reserves. They have a large amount of lipids that, in part, they transfer to their huge eggs (up to 4 mm in diameter). They lay few eggs (about 10,000–20,000 each) very rich in fats, which allows the future larvae to survive when the system is dormant (in winter). A slower metabolism also implies slower growth and a more progressive maturation. Icefish do not lay until they are 5–8 years old, and they can live for more than 20 years, which is a long time for a fish of modest size.

But when did these fish become different from other fish fauna? "The diversity of species in Antarctica has changed over the last forty million years," says Joseph Eastman of Ohio University; "that is, since the end of the Eocene, when the continent began its true isolation from the rest of the planet." As we have already seen, the water temperature was not always so low in this part of the planet. The most drastic change occurred mostly in the last 10 million years, at which time fish and other organisms had to "decide" whether to stay or flee from such extreme living conditions. "The fact that there are so few species is clear in the case of fish, their evolutionary radiation is more recent than that of other organisms," says Ian Johnston of the University of St. Andrews (Great Britain), "and living under conditions of extreme cold and strong oscillations of food is not easy." It was 38 to 25 million years ago that the way opened to the creation of these glycoproteins (antifreeze agents), as well as a whole series of adaptations to conditions completely unlike those of other fish on the planet, but it was not until 10 million years ago that their formation and metabolism were perfected.

Antifreeze substances have arisen independently in several species of organisms, but few groups have developed them as well as icefish. "The absence of competitors led to the occupation of a number of ecological niches that were exploited by icefish and a few other groups," stresses Johnston, "but we don't know why this was the 'chosen fish' group." The curious thing is that there are very few pelagic species, most of which are benthic. Why? This is another good question that is not easy to answer. Specialists argue for two compatible possibilities: the fact that this way less energy is consumed; or the squid's occupation of its ecological niche in the pelagic zone. In any case, these unique fish (no more than 1.3% of the species of the entire planet) are an important part of the system, feeding on krill and serving as food for a wide variety of predators.

This battery of adaptations to cold leads fish to a real problem: stenothermia. Icefish can live only between -1.8 °C and +6 °C, and beyond this temperature their physiological system fails and they die. At a temperature where tropical fish begin to feel the cold and die, 6 °C, icefish start to scorch. This narrow temperature range makes them highly vulnerable to any change that exceeds their tolerance in this part of the world and forces them to seek refuge in areas where temperatures remain stable.

A paradigmatic case is that of the pelagic silverfish, *Pleurogramma antarcticum*, a species of sardine that is the food base for many predators such as penguins (up to 80% of the Adélies' diet is based on this fusiform fish), seals and flying birds. In the northernmost part of the Peninsula, these predators no longer have this shiny armored fish in their diet. From being very abundant in the 1950s, it disappeared. By checking the temperature time series, scientists have realized that air temperatures had risen to 6.5 °C in the middle of summer, with repercussions on water temperatures, especially in the northernmost areas. So much for a species that barely survives in water above 2–3 °C. What may seem ridiculous to us may be the reason for the displacement for a species without any adaptation mechanism to withstand certain changes. In searching for *Pleurogramma*, the scientists saw that in just a few decades its populations had moved southward, to the ice margin far from their original distribution.

As we will see below, the peculiar characteristics of these fish (slow growth, very specific location, limited reproduction, limited tolerance to temperature changes) have led to the collapse of fishing grounds that were exploited earlier, especially in the Scotia Arc (between Patagonia and Antarctica). In South Georgia, the extraction of almost half a million tonnes in only 3 years destroyed the fishing grounds in that area, with recurrent exploitation without quotas (and, above all, with no knowledge of the species) of the various areas where vessels could penetrate with little risk. To date, 15 species of icefish species have been exploited and in no place have the stocks recovered, due to a synergistic effect whereby penguins, seals and cetaceans exploit the fish as a source of food. As krill have declined in some highly productive areas, the fish have suffered, with 50–60% of individuals developing a decreased ability to mature the gonads and thus perpetuate stocks. "Icefish populations are in decline," says Ryszard Traczyk of the University of Gdansk in Poland, "but it is not only climate change and overfishing that are affecting them: competition with other species for food, slow growth and lack of adequate food are also acting on the three species we studied." Synergistic effects (fishing, natural predation and lack of food prevents the normal development of eggs) are endangering the presence of this unique type of fauna on our planet.

10

Birds at the End of the World

Penguins: The Bird that Came in from the Cold

Waiting for me at the *Polarstern* heliport is the most senior pilot from the voyage that I made in 2003. I'm a bit puzzled, because in theory I don't get to fly, which is a pleasure offered only in dribs and drabs, and only the day before I had been up in the air helping to transport materials to Joachim Plötz's base in Atka Bay, the southernmost point of the Weddell that we are to visit.

I've packed my photographic equipment. Arriving at the hangar, I'm told to put on my 'hard flight' suit, the one that allows you to survive for 10–20 min in the water without freezing to death (otherwise, because of the cold, you usually don't last more than a couple of minutes). I am surprised, but change into the suit that even I can't fit. The pilot tells me to get onboard the aircraft. I obey, and see him manipulating something on my back. Apart from a conventional seatbelt, attached to me is a hook from the top of the cockpit. Before getting onboard himself, the pilot looks in at me and opens the door: I'm to fly with the cockpit open, so I can take photographs (Fig. 10.1).

I'm on Cloud Nine. Hardly anyone is allowed to do that. The gods and Wolf Arntz (the expedition manager) are on my side. We start the flight and I feel my blood pumping, and the adrenaline starts to kick in. It's going to be a completely different flight from others I've been lucky enough to take, I can feel it.

And so it is. We fly to a penguin colony, where hundreds of emperors, adults and young, are crowded together in the sun. It is a spectacle of the first magnitude, and the pilot gets as close as he can without scaring them, and I feel privileged. I take a series of photos that will never be worthy of National

© The Author(s), under exclusive license to Springer Nature Switzerland AG 2022
S. Rossi, *A Journey in Antarctica*, Springer Praxis Books, https://doi.org/10.1007/978-3-030-89492-4_10

Fig. 10.1 Emperor penguin colony, with adults and chicks

Geographic yet leave me satisfied. On the way back, the pilot throws us into a series of pirouettes, just for me. When we land, I give him a hug of thanks, knowing that it is likely that the occasion will never be repeated in my lifetime.

I had already seen penguins from the ship and near the bases, but never like this. They are clumsy animals on land yet extremely agile in the water. They are made to swim. And the ones that impress me the most, without a doubt, are the emperors. They are large (about 120 cm high, and can weigh more than 45 k), elegant (always with their head up, with that mixture of black and white tinged with a yellowish orange that gives them the appearance of being in a fashionable tailcoat) and hydrodynamic. They look like a bullet, ending in a sharp beak 10 cm in long, in some cases. They are often seen following the boat to see if anything falls overboard. They follow us because they have seen that we catch fish and discard some, although the truth is that we drop very little because of our everlasting greed for samples. I watch them swim and I notice how they move without difficulty, not like on the ice; there, they prefer to slide on their bellies rather than walk hesitantly. Their double layer of feathers, short but extremely thick, provides more than 90% of the protection they need from the cold and is impervious to water. Moreover, their layer of blubber (double that of any other penguin) makes them more adaptable to extreme conditions.

In fact, the emperor penguin works the other way around to other Antarctic creatures. While others flee the cold by migrating north, emperor penguins may travel more than a 100 km toward the south, grouping together inland. There, after the mating season, the males incubate a single egg for more than 60 days in June, when the light is almost extinguished at the end of the austral autumn. The females have moved north to feed, and the males must keep their temperature at 39 °C (and that of their egg), safe in the ferocious wind, which can reach over 150 km per hour, giving a temperature of below -40 °C.

How do they survive? This must have been the question asked by the three scientists, members of Scott's doomed expedition to the South Pole, who in the early 1900s discovered this unprecedented behavior in emperor penguins. Bowers, Wilson and Cherry-Garrard's hunt for eggs, later described by Cherry-Garrard in *The Worst Journey in the World*, observed the penguins trooping along far from the edge of the ice floe. This strategy evidently works for emperor penguins, and we have to understand how. To begin with, these birds have a large wingspan, so the ratio of their exposed surface area to volume is smaller than that of other penguins. They therefore lose less heat. This is also true because their feet are highly insulated from the ice and their limbs are shorter than in other species (from where a great deal of heat can be lost). They also have a more developed internal circulatory system than their closest relative, the king penguin, which is smaller and less able to withstand the cold. Their nasal passage is a marvel of heat recovery. Cold-adapted animals must not allow icy winds to enter directly into their bronchi and lungs, as this would be fatal. They also have that thick double layer of feathers, accompanied by a greater layer of fat than other penguins. But all this is not enough. They need something more. And that something more lies in their behavior. They crowd together in a circle, alternating almost mechanically between being inside and outside, in an endless rotation that allows them to withstand blizzards.

Into this equation we must factor their huge tolerance of their companions, as all their aggression is inhibited—imagine if they were unable to tolerate each other within just a few centimeters. That is why, whereas others have to flee to seek respite in more northerly areas, emperor penguins can be considered the true inhabitants of Antarctica, able to withstand its inclemency.

When the females return, the eggs have either already hatched or are about to do so. The females return fattened, stuffed with fat and food that they will use to feed their chicks. The males then take the same path back to go fishing. They will catch mainly *Pleurogramma*, a pelagic fish very common in these waters, although they do not spurn squid and krill. Their dives will be shallow, lasting about 2–3 min, but if necessary they will go down to more than 500 m, holding their breath for up to 20 min. They are adapted to this, as their dense, strong bones can withstand high pressures and their hemoglobin is, capable of capturing large amounts of oxygen (and carbon dioxide). Their metabolism will slow down, circulating less (as we will see in seals, too) through their extremities, through which much heat could be lost.

Watching penguins swim is, as I said, a wonder. You see long-lived animals that, given an easy life, can live to 50 years (only 1% reach that age), maturing late and raising a single chick each year. Not all chicks survive, and Antarctic petrels are responsible for killing more than 30% each season, along with other predators or scavengers controlling this population. Additional adverse effects on their population due to environmental change is gravely worrying.

Estimates are somewhat confusing because, like those of all Antarctic animals, the monitored populations of emperor penguins are found along scientific routes or near bases. "After more than a hundred years after their discovery," says Barbara Wienecke of the Australian Antarctic Division, "no exact population figure is known." At the 33 routinely monitored sites, about 270,000 to 350,000 individuals have been counted (2008 data), but it is possible that there are many more. Considering that they eat 1–2 k of fish each day, their impact as predators is not insignificant (it could be a minimum of 200 hundred thousand tonnes of fish per annum, constituting up to 80% of their diet). If there were more penguins, the figure would increase. But in some cases there are fewer. It has been proven that over 50 years in Terre Adélie, 50% of the emperor penguin population has disappeared. This extended time series was made possible by the presence of a nearby base that was able to monitor the fluctuations in the numbers of adults, juveniles and hatchlings. Mortality increased as water temperature increased, and especially as ice cover decreased in the area. "Somewhat paradoxical," notes Christophe Barbraud of the French CNRS, "because the less ice there is, the sooner they reach their spawning grounds; but the less ice cover, the less production, copepods, krill, fish, squid …".

Penguins face several problems, especially in the Antarctic Peninsula. That's where there is the most variation, and any unexpected factor may help to throw their survival plans out of whack. "Macaroni or Adélie penguins are used to some change in diet, but, like other species, if the changes are too rapid, they may not be able to keep up," says Robert Pitman of the National Fisheries Research Service in California, Unite States. In this case, climate change, as we shall see, has much to do with it, but so does commercial krill fishing. "Overfishing of krill could happen at any time," Pitman adds, "because the processing of krill shelling, which was a big problem, has been perfected and is too interesting a resource in a world of overfishing."

This fact could greatly affect these two species and, of course, the many other links in the chain that depend on krill, as we have seen. In the case of Macaroni penguins, 20% less in their diet would mean their having to swim further for less, thus they would lose the ability to keep the next generation alive. "There is a point after which they cannot catch enough food for themselves and their young," adds Dr. Cresswell of the University of California at a recent SCAR conference; "make no mistake, all animals adapt, but extremes are very hard and these animals will endure fluctuations to a point of no return."

Commercial krill fishing (and of other elements, such as *Pleurogramma*) takes place in summer, the season when all species need maximum food for their own survival at shallow depths in coastal areas, where more penguins and other animals depend on the decapod manna that forms swarms that are kilometers long. But, according to Pitman, there could be an additional problem that is, as yet, little studied. Having to swim further to catch their prey and with less ice cover may leave the penguins more exposed to predators such as killer whales and leopard seals. As always, it is a complex situation in which there is more than one factor acting as a tipping point for the local disappearance of these (and other organisms). Just think that in certain areas the records show that more than 80% of the Adélie penguin population has disappeared: this is another species (like the emperor) that is not very fertile, matures late and is highly vulnerable to major changes. These extreme-cold specialists will soon see their range of action restricted, and we will witness just pockets here and there where those who manage to find a suitable refuge attempt to survive.

Fliers

From his privileged platform in the spacious cockpit of the oceanographic icebreaker *Polarstern*, Henri Robert of the Royal Belgian Institute of Natural Science gazes through his binoculars at an apparently monotonous landscape: the ice floe. In front of him, kilometers of water and ice stretch out, barren, without apparent life. But if we look closely, several species of birds follow the boat or move close to it, flying without apparent rest. They are petrels, albatrosses and other birds that live in or around this part of the planet, tirelessly searching for food for themselves or their offspring. Henri's task is to note their presence, count their numbers, locate them in specific areas and, on his return from this voyage of almost two and a half months, report the data that will help us to understand better how little is known about a these species that seem to have been born to defy our planet's most adverse conditions (Fig. 10.2).

Fig. 10.2 The birds may travel thousands of kilometers in search of food

If we dive into the specialized literature, we see that little is known about most of the 35 species of flying birds that populate the southernmost seas of the planet. To begin with, almost nothing is known about their populations—how many there are, where they are, how they move. Specialists on the subject are scarce, and the data are scientific interest rather than romantic. If little is already known about other species such as seals or penguins, the count of a hundred million birds around the white continent is a speculative estimate that the authors themselves admit lacks any solid basis.

Of the southern fulmar (*Fulmarus glacialoides*), a minimum population of 400,000 is estimated in the most-studied sector, the area around the Antarctic Peninsula. This estimate rises to more than a million if we begin to speculate on areas similar to those studied but which are less accessible or visited. It is known that more than 70% of this species is centered on the Scotia Arc, yet all specialists recognize that there must be several million and that very little is known about areas such as the Ross Sea or east of the Peninsula, where they are also known to live. Undertaking a census of any wild species is difficult, but counting Antarctic wanderers is a real challenge.

Sometimes we find high densities in the open sea, or near the ice floe, but their nests may be far away. Counting birds may seem an easy task, but it is complicated to avoid making mistakes by repeating a count, and it relies on locating the individuals well and getting as close as possible to a credible figure. The migrations of these animals are long, the areas tracked by individuals in search of food very wide. Coastal species (such as skuas and cormorants) do not have such an imprecise and wide range (although skuas have been found in the middle of the continent, thousands of kilometers away from the coast); however, species termed "pelagic" (because they feed in the open sea or near the ice floe, sometimes thousands of kilometers away from the mainland) can be found in the most unexpected places in search of squid, fish or krill.

But they all return. In one way or another, all flying birds all have to nest, so their wanderings on the open sea, which can last for months, come to an end at the time of procreation. That is why the areas where the most accurate counts to be carried out are precisely those places where the land provides a place for them to lay, in the middle of so much ice. And, outside the Antarctic Peninsula and the Scotia Arc, even in summer such places are scarce. Finding a suitable laying site is tough, and the struggle, the rivalry between the contenders is intense. Time is against them, because the window of opportunity is short. The chicks have to grow in record time compared to similar species in warmer environments, because the imperative of the ice does not allow any hesitancy. The ice floe and the available food soon disappear, and the next generation has to be ready to move. But what happens when, in addition, they encounter a changing climate?

Consequences of Climate Change and Fisheries

At the Antarctic bases where they have been working for decades, scientists have been able to observe changes in the populations of some species that may relate to temperature variations. After more than 55 years of monitoring bird populations, Christophe Barbraud and his colleagues have found that the birds, contrary to behavior noted at the North Pole, are arriving to nest later and later. "About 9 days on average depending on species and area over the last 50 years," explains Barbraud; "they also lay their eggs about 2 days later than in the 1950s."

So what. What real change has there been? In these species, as mentioned above, the period for laying is short—shorter than in other species. A change of days in some of them can, in a sense, be compared to a delay of weeks in warmer parts of the world with a wider time window for feeding. This delay of 9 days may fail to allow the chicks to mature in time, as under the current system they are already sufficiently stressed. "If there is a real mismatch between available food and nesting times, we may be facing a real long-term problem for these animals," concludes Barbraud.

In South Georgia, the albatross population has declined by 40–60% in 35 years. "Albatrosses, like other oceanic birds, feed in a very changing sea," says Deborah Pardo of the British Antarctic Survey; "food shortages due to fishing and climate change-related phenomena are decimating them. The El Niño phenomenon is an example of a disruption in their lives because of the changes in the food chain that it produces." The same happens with Adélie penguins, which in some years may see entire broods die in some areas due to heavy rains in places where it never rains or a lack of sufficient food due to the ice dynamics. However, experts like Barbraud agree that they see much slower changes in the south than in the Arctic.

Using stochastic models to compare changes in the distribution of circumpolar birds over the next 50 years, Barbraud and his colleagues have found that not only the laying of eggs but the distribution of food are at risk for many species. Species living further south will be the most affected by both temperatures and ice cover. The model, however, lacks a basic factor: birds' ability to adapt to the new conditions. Snow petrels feed in the polynyas, which indirectly depend on the production of the microscopic algae to feed the krill and other pelagic organisms living in association with the ice dynamics. When these dynamics change, the birds will have to adapt, move or perish. "Any change in the phytoplankton will affect the birds, even if they don't feed directly on it," insists Barbraud.

In fact, birds are considered a good indicator of variations in the Southern Ocean. A long time series (from 1952 to 1999) carried out in Terre Adélie shows that the giant petrel drastically changed its presence in the island's environment (an average of almost 4% fewer each year), yet that of other birds hardly changed at all or even increased their numbers (up to 3.5% more per year, in the case of skuas). However, controversy arises when we look for the factors behind the population fluctuations. Skuas increased at the same rate as Adélie penguins in the same area, and were directly responsible for regulating the population of these wingless birds by eating their eggs and chicks. An increase in penguins may be related to increased food availability due to changes in the quantities of krill and other pelagic organisms in the island environment, most likely the result of declining whale populations.

So, what influenced the decline of giant petrels in this area? It may be that the presence of man is directly responsible. Indeed, the species whose numbers have declined in recent decades in this area are the rarest and most elusive: the giant petrel and the emperor penguin. Neither are fond of being close to human bases or settlements, so their decline may be related more to our direct action than to climate change. Helicopters, noise and movement do not help at all to maintain the populations, whose fewer individuals are increasingly isolated.

More Palpable Dangers

Other much more direct but less visible factors affect these indefatigable birds. "Pelagic birds and penguins are the most affected by by-catch," reveals Henri Weimerskirch of France's CNR, at an Engineering in Medicine and Biology Society congress: "The means of protection are clearly insufficient." For example, of the 18 existing penguin species, 14 have been found hooked and drowned in industrial fishing lines. All ornithologists agree that fishing is a huge threat, and is spreading more and more throughout the waters of the Southern Ocean. The long, baited lines to catch all kinds of fish are a death trap for these birds, which can travel more than 3500 km only to meet their death in the stupidest possible way: by a longline hook. Unfortunately, there are no precise numbers—in fact, there are no numbers at all. The boats that fish with this type of gear are tired of pulling birds out of the sea, and only now are some beginning to count their accidental catches. "The problem is that no real sanctuaries can be created, because there is no fishing in the high Antarctic, but there is fishing in South Africa or Patagonia, and much of the fishing there is uncontrolled because it is done in international waters or with very lax control measures," says Henri Weimerskirch. On a single fishing line, up to 36 dead birds of a single species have been detected. That is a lot.

But it is not the only problem that the birds face. Even in and around remote Antarctica, for many species plastic is synonymous with death. Jan van Franeker and Phil Bell (between 1985 and 1987) found that no less than 75% of storm petrels had plastic in their stomach. The plastic drifts on the water, brightly colored or white, attracting these avid, hungry birds that swarm over the most remote places in search of food. They see a shred of plastic, dive in and eat it, thinking it is a fish, a squid, a crustacean … something edible. And then, if they survive, they regurgitate it to their chicks hundreds of miles away. "In Antarctic waters there is very little plastic," Franeker continues, "but in the overwintering sites further north, they come from everywhere."

The circumpolar current slows down the entry of plastics, as well as the fact that they are distant from the main sources of mass release (the coasts of anthropized areas), but plastic has still become part of the birds' diet. The related problems are several: suffocation and toxicity are the most obvious, but there is a third, more subtle problem for chicks. "Birds eat something that doesn't bring them any food," says Bell, "but fills their stomachs; when they come in and regurgitate, some of the food is not food at all, it's simply a piece of plastic with no nutritional value." This is also true for adults, whose food reserves are depleted by a false intake. Not all species are affected, and it is mostly those that feed in specific areas, such as around outcrops or along shipping lanes, where large quantities of waste are dumped into the open sea.

As I watch one of the birds that has been following us for days (a grayheaded albatross), I think of his journey, of everywhere that he has had to fly, of how vulnerable I am in the face of the oceanic nothingness while he glides unperturbed. I think of the winds he takes advantage of to reach the upwelling places, where he concentrates on searching for food, and how far he will be from the nearest landing place or platform on which to rest without the lurching waves. And I think of the fragility of his condition as an adventurer of the sea, now afflicted from several sides by a series of direct and indirect disturbances that will make his life a little more difficult.

11

Large Mammals of the White Continent

Why Do Whales Go to Antarctica to Feed?

We're all holding our breath. We've been told that whales have been sighted from the icebreaker *Polarstern*'s cockpit. This is a huge space that we can all access unless there are restricted maneuvers or the captain feels that it would better for his officers to have a calmer environment at a potentially dangerous juncture. No one speaks. Everyone is watching, with binoculars or telephoto lenses or straining their eyes to a horizon littered with icebergs in an ice-covered sea. The sun is low, so the light is perfect: hard, orange, full of nuance.

In 2000 on my first expedition, the official bird spotter (Hans) was always the first to see any whales. Calmly, in that tense atmosphere of waiting, he has now pointed out on the horizon the jets spurting from the blowholes of seven or eight humpback whales that he has glimpsed in the distance. The captain turns toward that area, to try to get closer. We are all excited about the possibility of seeing whales searching for holes in the ice to breathe. And finally I see them. The jet of water and air shoots out, forming a sprayed curtain of a yellowish color due to the twilight sun. It is a magnificent spectacle that I savor with delight because it is possible that I will not see them again, even if the expedition takes 2 months. It is not easy to see whales, even in Antarctica, a place where the abundance of food attracts them year after year from distant oceans.

Whales migrate to the Southern Ocean because of the abundance of food. Krill and small fish (*Pleurogramma*) are the food of the mysticetes whales (toothless, with baleen hairs that have evolved to hold small prey). The high

© The Author(s), under exclusive license to Springer Nature Switzerland AG 2022
S. Rossi, *A Journey in Antarctica*, Springer Praxis Books,
https://doi.org/10.1007/978-3-030-89492-4_11

concentrations of these prey attract the giants here to feed and thus accumulate fat, in one of the largest oceanic gardens on the planet.

Rather than here, they will mate and give birth in warmer waters in more suitable areas, where food is scarcer yet predators are also less abundant. Because Antarctica is home to other whales, too—the odontocetes (whales with teeth), some of which are whale killers (and not necessarily 'killer whales', as we are used to hearing). They follow the three species of mysticetes, especially the small minke whales. "There is a strong correlation between the distribution of minke whales and Type A killer whales, those that eat whales," says Peter Corkeron of James Cook University in Australia. He and his colleagues have observed that where these small mysticetes are found are higher numbers of this type of whale, and they have also noted that they coincide with blue whales. "In this case, there doesn't seem to be predation in this water by killer whales on blue whales, they are too big," says Corkeron; "the coincidence of minke and blue whales is because they exploit the same type of prey: krill". They may compete for this food resource, and the greater numbers of minke whales (even though smaller) may be a factor in blue whales' slow recovery.

Killer whales can also eat krill, but only Class C killer whales, which feed mostly on fish. "Class B killer whales," says Corkeron, "are somewhere in between, eating penguins and seals." They form three different morphotypes (by size, skin spots and shape), yet there is still doubt as to whether they are distinct species. Type C killer whales are the smallest, and Type A the largest and most aggressive. They can be found individually or in groups of up to 30 individuals, mostly in open sea areas, although the fish-eating ones go deeper into the ice floe, where the capture is safer.

No whale is unique to the white continent. The Arctic has its beluga whales and narwhals, which are found only in these northern waters, yet the Antarctic has no unique species—they are all always in passage. Whales are dependent on the topography of the bottom, the upwelling areas, the dynamics of the ice, algae and krill. And where there are submarine canyons or mountains in the middle of the continental shelf it is easy to find krill, as such areas have the perfect productivity characteristics to support the large schools needed to feed fish such as *Pleurogramma*. They all play a role in the matter and energy cycles of these polar systems, a role we tend to forget.

Whales follow the food, like all animals. Krill is abundant near the ice, so they go where the food is, and that means they are increasingly being found further south. "Humpback whales routinely migrate south to fatten up," says Susan Bengston-Nash of Australia's Griffith University; "they are sentinels of the changes occurring in these latitudes; they have less and less blubber, probably because the abundance of food can be scarce in certain areas." Not surprisingly, their prey is suffering not only from climate change but from commercial fishing. Resources are dwindling and, although some whales appear to be largely unaffected by the changes, others may not recover by 2100 to levels similar to those before their "ecological" near-extinction in the late-nineteenth and early-twentieth centuries. But why are whales so important?

The Real Role of Whales in the Carbon and Nitrogen Cycles

Our anthropocentric vision prevents us from seeing the importance of our direct competitors for the planet's living resources. Along with other animals and plants, we have displaced whales from their role of controlling food webs, nutrient cycles and even the production of the planet's living resources.

Impossible? At their current numbers, whales continue to play an important role in ecosystems. In the Gulf of Maine (NE United States) a group of scientists have made a few calculations. "Whale feces are a potent fertilizer that stimulates the productivity of microscopic algae," says Joe Roman of the University of Vermont (United States). Here alone, the few whales monitored by a network of scouts demonstrate that, along with the few seals that remain, they can put into circulation some 23,000 tonnes of nitrogen, more than all the rivers in the area combined.

That's a lot of nitrogen, and it's second only to groundwater and other waterways that flow into this area from the land. "Before they were reduced to nothing, the whales could surely put up, along with other large vertebrates, as much or more nitrogen than is transferred from the atmosphere," Roman concludes. Even if these calculations seem exaggerated, the biomass of some three or four hundred million whales had to have exerted an influence on the production of our oceans. Moreover, the whales would have been able to immobilize an enormous amount of carbon in their long-lived bodies. They served as fundamental regulators of our planet's biogeochemical cycles.

We must take these contributions into account. It is vital to explore how the simplification of global systems by the systematic destruction of the planet's complex forms (large trees, whales, tuna, deep corals) has prevented these systems from retaining carbon, moving nutrients and transporting matter from one place to another in our marine ecosystems. In fact, whales can capture their prey at great depths in some cases, acting as true matter elevators from the depths to the surface. They defecate, then the dissolved matter is taken up by phytoplankton, which is then eaten by zooplankton, which in turn is eaten by small fish or krill (which also feed on algae) and so moves to the depths where the whales consume it, starting the cycle all over again. This carbon sequestration has been tested in several places on the planet with other large animals, such as the fauna of the Serengeti. There, it is the wildebeest, antelopes and elephants that act as modulators of fertilization and grassland productivity. They are also responsible for slowing down carbon cycles by immobilizing part of it in their body structures.

And in Antarctica? Here, whales transfer carbon and nitrogen from the bottom of the food chain to the top almost directly, through krill. It has been shown that the decline in krill has coincided, in part, with the decline in the number of whales in this part of the world. Fewer whales, less feces, less production by algae, because part of the nutrients are not reintroduced. It is a possibility to consider, although of course it is not the only one (remember the ice fluctuations).

What is logical is that krill dynamics influence whale dynamics. It must be remembered that there are species of mysticetes that feed almost exclusively on krill and small pelagic fish. Even with numbers a pale reflection of what they once were, mysticetes whales may be consuming 4–6% of Antarctic krill. And what they give back in the form of feces is not just nitrogen. It is phosphorus, iron (essential for microalgae, as we have seen before), micronutrients …. Could primary production (phytoplankton) in the past have been higher due to the action of whales? In the last century, after the disappearance of whales from the white continent (and elsewhere), it was found that in 80% of the sites studied primary production had decreased. On a global scale, large mammals (and large sharks and fish) provide a service to the ecosystem that is essential to its proper functioning. They may even influence climate regulation through carbon sequestration in their large and long-lived body structures and by increasing the productivity of algae that capture CO_2 for photosynthesis. And Antarctica is one of the most productive places on the planet, where many whales went to feed in the past.

Recovering from the Catastrophe

In March 2011, while standing on the stern deck of the *Polarstern*, I saw a whale breach. Near to the boat, a humpback suddenly came out of the water and lay on its back. Several of my colleagues were on the same deck, but I was the only one in the group who saw it and, since it did not come up again, my colleagues were skeptical about my prodigious vision. Fortunately, Philippe, the second-in-command, had photographed it from the cockpit and at the daily photo show in the briefing room we were all able to laugh about my ghost whale.

Seeing that spectacle (similar to the one I described at the beginning of this chapter, but closer) made me reflect on whether we are treating whales well after centuries of hunting, in the case of some species decimating them to near-extinction. I have tried to explain their real role with a series of examples that are increasingly being taken seriously by ecologists and people trying to understand the profound changes that we have been perpetrating over decades, centuries, millennia of interference in marine ecosystems. After the 1980s moratorium on whaling, whale populations should be recovering, especially in an ocean as distant and pristine as the Southern Ocean. Is this the case?

Even if they are recovering from the spectacular decline in Antarctica (and all over the planet) caused by relentless hunting, whales continue to have problems here. The first official whaling expedition was made by a Norwegian ship that returned in 1904 with 195 dismembered specimens. Between 1910 and 1930, a large-scale whaling industry was created in the area, displaced from fishing grounds where there was little chance of finding so many animals. The efforts of cetacean hunters increased 30-fold, reaching a peak of 40,000 specimens in 1931. As early as 1925, over-exploitation of whales was recognized by experts yet, just for a change, this was to no avail. Before the Second World War (when whaling stopped almost completely), the numbers caught reached 45,000 in 1938. It was an extremely lucrative business that provided meat, oils and raw material for countless manufactures. In 1963, protection treaties began, after realizing that in certain areas some species were almost extinct. In 1987 a treaty was created for their total protection, with strong complaints from countries such as Japan, Russia or Norway, which saw their business in polar (and non-polar) waters disappear. In only 70 years more than 1.3 million specimens had been hunted in the Southern Ocean (official data).

Little by little, some species are recovering. The species that is doing the best is the minke whale, a small mysticete with between 600,000 and 700,000 specimens, while humpback whales have about 80,000 specimens, according to the disparate (and always complicated) censuses by specialists in this area of the planet. The blue whale is not so fortunate. There are about 10,000 individuals and the numbers do not seem to be recovering at the expected rate. Its stock is estimated at less than 1% of what it was before its ruthless persecution. This is not surprising. According to twentieth-century statistics, between 20,000 and 30,000 specimens were hunted every year.

The moratorium and close surveillance have led to a considerable decrease in whaling today, but it is still ongoing. In such a remote part of the planet, control is difficult and only a handful of madmen (the Sea Shepherd organization) try (very aggressively) to prevent poaching or scientific hunting (see below). "I have not gone all these years to sea to simply witness the atrocities committed by whalers on the most intelligent beings in the ocean," Paul Watson, co-founder of Greenpeace, assures us: "We are policemen of the sea operating legally under the guidelines of the United Nations World Charter for Nature." Such a visceral reaction, involving the sinking of several whaling ships in the area, comes after indiscriminate hunting and little desire to change things on the part of the whaling industry, which is pushing to reopen the ban on the minke whale, at least, which they say is possible to exploit sustainably.

It is possible that, despite not being hunted as in the past, populations of certain species are still headed for inexorable extinction. The competition with minke whales for food (krill) could be behind the unpromising recovery data on blue whales, the largest animal that has ever existed on Earth (or rather, on the globe). "The models predict an uncertain future for the different species of whales," warns Vivitskaia Tulloch of the Australian University of Queensland; "while humpback whales could recover, blue whales could have a much more difficult time." Changes in krill (which vary from area to area, like everything else in the vast Antarctic) will affect the competition for food between several species of animals that consume large quantities of this crustacean and depend on it for their survival. Climate change, ocean productivity, pollutants and shipping traffic will continue to deplete a group of animals that migrate long distances to build up reserves. The latest danger is the growing tourism involving apparently harmless sightings in the Antarctic Peninsula, which can actually alter whales' feeding routes and create specific local problems that add to the global issues. Unfortunately, the lack of data in such a heterogeneous area of the planet does not facilitate the management of these animals, which seem fundamental to the functioning of the Southern Ocean in particular, and the cycles of carbon and other elements in the seas of the entire planet.

Bidding for an Illegal Hunt?

An article in a prestigious polar scientific journal (Polar Biology) suggests that in sub-Antarctic waters minke whales (one of the smallest cetacean species) have lost 9% of their body fat over the past 18 years due to the greater scarcity of krill. "The increase in predators of this crustacean and the decrease in its biomass are possible causes in the decline of reserves," writes Kenji Konishi from the Institute of Cetacean Research in Tokyo in collaboration with the Institute of Medical Science at the University of Oslo: "They may have lost up to 0.02 cm per year."

A further paper, by Susan Bengston-Nash, has come to the same conclusion, yet there is a big difference in the methods used. In fact, there is controversy over what the Polar Biology scientists claimed was necessary to establish an adequate sample size and the type of sample used to reach these conclusions. "To avoid statistical errors, 2890 mature male whales and 1814 pregnant females were hunted," Konishi and colleagues state in their article. In fact, more than 4500 minke whales were caught between 1988 and 2005, increasing the numbers hunted from about 250 to more than 400 each year. Bengston-Nash and her colleagues did not use these methods, as they were entirely unnecessary.

Some scientists are deeply concerned about this paper by Konishi and colleagues. It might seem like just one more article but, in the words of Research Professor Josep Maria Gili, a very serious precedent has been set: "The journal *Polar Biology* is the most prestigious polar journal of marine biology in the world. By accepting this article, two serious mistakes are committed: the first is to tender whaling, which is illegal, through a work that appears in a rigorous scientific journal; secondly, it is telling you that anything goes in science to reach a specific goal." As co-editor of this journal, Professor Gili is concerned about the possible consequences of this article: "It is as if now it were said that, to study what seals or penguins eat, we need to sacrifice five or six thousand specimens in pursuit of a scientific article Where is the limit? Where are the ethics?"

What is disconcerting is that the other scientific co-editors did not raise any formal complaint to the then editor-in-chief (the article was published in 2008), Gotthilf Hempel. "I am seriously rethinking my collaboration in this polar science outlet," added Gili at the time. For Professor Hempel, however, the article is perfectly appropriate: "We sent it to two prestigious scientists and they saw no errors in the approach or in the way of treating the data." Any scientific work must first pass through referees, two or more scientists anonymous to the author of the work, who give their opinion, suggest corrections or even reject an article on the basis of the rigor of the data and the approach taken. "In addition," adds Hempel, "the minke whale is not an endangered species. It is estimated that it has a population of more than 500,000 specimens around the planet." It's an old controversy. "As the work has been published in an internationally recognized scientific journal, the Japanese and Norwegians have a tool to defend their covert hunting policy," continues Gili.

The minke whale (*Balenoptera acutorostrata*) is one of the most hunted cetaceans on the planet. According to the Convention on International Trade in Endangered Species (CITES), by not being an endangered species this whale is at the limit of exploitation, and the evolution of its populations must be closely monitored. In 2007, the Japanese government had intended to catch 935 specimens, but the actions of conservationist groups caused this quota to drop to 551. "We had to stop catching 45% of what was planned," comments a representative of the Japanese Fisheries Agency. "We do not have enough time for research due to the assaults and impediments caused by conservationists." According to the Japanese, their observations are crucial to perceive upward or downward trends in the populations of whales. They strongly criticize what, for them, is Europe's hypocritical position regarding the whale problem. "Many European countries exterminated whales between the late nineteenth and early twentieth century and now they prevent us from hunting them in a sustainable way," says Shigeki Takaya from the same Institute of Cetacean Research in Tokyo. Whales processed by scientists pass into the hands of the same company that funds the Institute itself and markets their meat. "Over the last few years there has been a marked increase in catches," says Junichi Sato of Greenpeace.

Conservationists are very concerned about the upward trend, especially in Antarctic waters. "The Antarctic Treaty specifies that fishing in its waters is prohibited for these and other species," adds Sato. And the targets are shifting to other species: "This year, fifty humpback whales are planned to be hunted in the waters of the white continent," says Geoffrey Palmer of the New Zealand whaling committee; "some countries have already established zones in which we prohibit the hunting of any type of these cetaceans." These are countries like Australia, which has declared a zone of 200 miles (about 370 km) around its coasts in which there can be no hunting for these animals, considering it an inviolable sanctuary. The controversy continues, and several countries make the criticism that, despite the huge commercial operation and the large number of cetaceans hunted, the number of scientific works has been meager and questionable (fewer than 60 articles in over 20 years).

The Japanese have a long tradition of whaling, but it increased dramatically after the Second World War. At that time, whales were a cheap source of protein and fat for Japan (and other countries) and, although already active, whaling accelerated. Its decline came a few decades later. Japan alone went from a peak catch of 226,000 tonnes in 1962 to 15,000 tonnes in 1985 (just before whaling was banned worldwide). The declining stocks indicated the imperative to stop catching them. The Japanese insist on the need to revise the treaties that ban whaling, arguing that there are now instruments to do it in a more sustainable way.

Seals: From Shy Crabeater to Fearsome Leopard

The other large mammals that live between the inclement icy surface and the bounty of the Antarctic waters are seals and elephant seals. There are not many species, but each has a specific diet, hunting depth and distribution that make it compatible with the others. The most abundant is the crabeater seal, whose diet (as its name indicates) is krill, which can make up more than 95% of its food. They are somewhat shy creatures, numbering in the hundreds of thousands (always difficult to count!) and consuming some 60 million tonnes of krill each year. This figure seemed to me to be an exaggeration, but Lloyd Lowry of the Alaska Department of Fisheries insists, "the estimates, made on a population of nearly a million seals, far exceed what is consumed by whales." There is no doubt that this species depends on krill biomass, like other species, such as penguins, fish, squid … all of them feed on the poor euphausiid. "We've found as many as 10,000 crabs about 4–6 cm long in a single stomach," says Lowry.

I have seen dozens of these crabeater seals looking at us, astonished, from their icy platforms, wary but with a certain curiosity to see the immense steel mass of the ship advancing on their isolated world, as though seeing something like that might happen just once in a lifetime. Some do not leave the ice floe until the ship is just a few meters away, breaking the ice under its powerful keel. Seeing them by the dozens, I understand that very few ships on the planet are capable of reaching so far out onto the ice floe, especially in the East Weddell Sea, where only three or four oceanographic vessels are able to penetrate (Fig. 11.1).

After these seals come Weddell seals, which are much fatter and capable of deeper dives. They can eat krill, but their main prey is the pelagic fish *Pleurogramma* (which can make up 90% of the biomass in certain areas). While crabeater seals do not usually dive to more than 200 m for food, Weddell seals go down to 600 m or more. During the day they forage for fish on the continental shelf, which, as we have seen, is very deep here. At night, somewhat less active, they search for prey only in the pelagic zone, not venturing deeper than 150–200 m.

Ross seals, their relatives in the Ross Sea, also feed on fish and squid, not descending far (usually between 100 and 300 m, although they can dive to

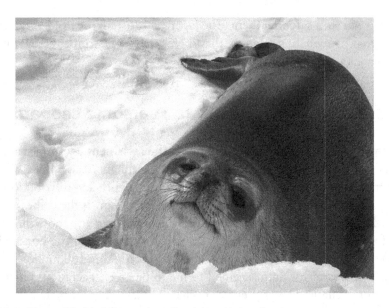

Fig. 11.1 Young Weddell Sea seal on the Antarctic ice sheet

almost 800 m) but making much longer trips offshore. In contrast to the Weddell seal, which stays close to the ice floe, the Ross seal travels up to 2000 km in search of prey offshore. For this reason, when it was discovered in 1840 and subsequently in several expeditions, it was thought to be scarce. The truth is that they were not seen because they were far from the ice, continuously roaming, making up to 200 dives a day of about 5–15 min depending on the time of year and the abundance of prey.

Although we may have an image of seals resting immobile on the ice, the reality is quite different. For example, elephant seals, absolute champions of deep diving among the pinnipeds of the area, spend no less than 80% of their time at sea and only 20% on the ice and land. Males spend a little more time lying in the sun, as they need a little less food than the females, who have to produce milk for their pups, but they dive to more than 2000 m to look for prey. One dive in ten goes deeper than 700 m, searching for fish and squid in the depths of the Southern Ocean. Both males and females can spend 20–30 min underwater, active and restless in search of their prey.

How on Earth do seals do it? These animals have several adaptations, present to a greater or lesser extent depending on their needs. The first is, of course, a thick skin and layer of fat to protect them from the cold and allowing them to consume energy at the right times. Their blood has much more hemoglobin than ours, allowing them to concentrate more oxygen, which will travel through their body distributing the precious gas to the cells and remove the carbon dioxide produced by respiration.

But that's not all. Seals are able to lower their heart rate (bradycardia) and cut off the blood flow to their extremities, concentrating it in the brain and central part of the body. Neuronal tissues are prepared for hypoxia (lack of oxygen) and high pressure. In fact, they are also prepared against collapse of their lungs: in the case of elephant seals, their body can take a pressure of more than 200 atmospheres, something impossible for a human being, no matter how well prepared (we have managed more than 600 m with special air mixtures, but working even at 200 m is a feat achieved by only a few).

The challenge is to adapt to the available food: to be able to go down to wherever it can be found. If the seals are near continental shelf areas, the dives will be at fairly fixed depths, but if they are in the open ocean they will follow the vertical migratory rhythms of zooplankton, the main food of fish and squid. "We err only on the side of following the movements of krill to understand the vertical or horizontal movements of seals," says Martin Biuw of the Norwegian Polar Institute; "copepods have a lot to do with it, because fish and krill themselves (which can eat small copepods) feed on these small crustaceans." Several species of seals have adapted to deep diving, thus avoiding competition for the (vertical) space with penguins and flying birds.

The last of the seals is the main predator: the leopard seal. The sight is impressive. Large, with an elongated head and a very wide mouth, it has been the protagonist of many incidents involving even human beings. In July 2003, a leopard seal dragged a young female scientist to the bottom of the ice during a routine visit to the frozen ice floe while she was freediving at the English base of Rothera. The news spread like wildfire, especially among scientists working on Antarctic issues. It had a special impact on me because just a few months later I embarked on the *Polarstern* on my second expedition to the Weddell Sea. Obviously, protected by a ship 118 m in length, the feeling is quite different, but the paranoia reached such a height that in a later expedition the scientists could not do anything for the safety consciousness of the German scientific divers. Their protocol involved not entering the water if a leopard seal had been seen within 24 h of a dive. The entire expedition involved collecting samples from beneath the ice floe, and the presence of these seals was continuous. To no avail did the seal specialists insist that the attack on Kirsty Brown at Rothera was a one-off occurrence—leopard seals may be aggressive, but they do not generally attack humans—nor the fact that the sailors improvised a metal cage to protect the divers. The entire expedition was wasted, at a cost of millions of euros.

It is true that the leopard seal is a fearsome predator. Its mere presence scares away penguins and seals, on whose sides one can see the evidence of its attacks. But the paranoia of the expedition's chief diver was excessive. Leopard seals have only one competitor to fear (apart from man, who erodes their ecosystem): the killer whale, especially Type A, which will not hesitate to chase even this fearsome seal.

What Happens to Seals in Winter?

In winter, the ice floe grows. Isolation can mean death from the intense cold and lack of food. The solution is to migrate, following the ice edge northward. "The animals continue northward, sometimes moving more than 700 km from the edge of the continental margin," says Joachim Plötz of the Alfred Wegener Institute. It is not a uniform migration, as it is necessary to follow the krill, the food that remains less active under the ice. Some species such as elephant seals undertake major migrations to places like Bouvet Island, thousands of kilometers from the continental margin, to spend the winter. Weddell seals are more conservative, following the "fast-ice" but never leaving the safety of the continental shelf. "They are perhaps the vaguest example of seals in this sense," says Plötz. They go into the open ocean but always return to the ice. In fact, they are the most accessible seals, and the easiest to study thanks to their more localized movements in space. At the edge of the ice floe they meet the penguins (apart from the emperors), which are also looking for a place away from the harshness of the Antarctic winter.

12

Climate Change: Not So Isolated

Climate Change in Antarctica

In 1995, the film *Waterworld* was released, starring Kevin Costner as the good guy and Denis Hooper as the bad guy. I went to see it but, as soon as it started, I knew that I was not going to like it: "In the near future, the melting of the polar ice caps causes the flooding of practically the entire planet …", an apocalyptic voice intoned, portraying a planet covered by the sea, leaving only some remote islands representing the Himalayan mountain range. A sovereign imbecility.

I feel that someone should advise these people before they make another million-dollar movie based on something so totally impossible and absurd. If the glaciers of Antarctica (the entire Antarctic) and Greenland all melted, sea level would rise a maximum of 70 m. That's a lot, and it happened in the Eocene, but mathematics tells us that it is far below, for example, Tibidabo mountain in Barcelona (500 m above sea level), let alone the Pyrenees, Alps or Rocky Mountains. What bothered me the most is that nobody gave any importance to the data, and the scene was presented as plausible. And that made me uneasy, a fear that has grown with age. Disinformation seems to be our daily bread, and in respect to climate change we seem to be reaching new record highs.

There is climate change. Much of it is undoubtedly of anthropogenic origin. And it has consequences, and we are already seeing and suffering them in countless places around the world. One of them, one of the most rapidly changing, is Antarctica. But if we use *Waterworld* as an example, we are going wrong.

© The Author(s), under exclusive license to Springer Nature Switzerland AG 2022   **91**
S. Rossi, *A Journey in Antarctica*, Springer Praxis Books,
https://doi.org/10.1007/978-3-030-89492-4_12

I like science-fiction movies, although I do require them not to insult my intelligence and to use a minimum of basic knowledge. That same movie could have been much better if the makers were properly informed, and the version that we are living today is much more real and will influence our lives much more than we may think. There are many uncertainties, which gives rise to direct discourse on the various problems that are gradually being brought to the fore. That is why it is difficult for natural, social and economic scientists to put things in order and present things in a linear way, to understand and make people understand what is happening, where we are going and how we are going to solve it. Nowadays it seems that the message has got through, but there is a lack of reactivity. We need to understand that this is something where we cannot resort to procrastination.

I remember very well a television news program in 1989 where the first IPCC-type report was presented, and I was very surprised. I was just starting college then, and it seemed far away. Now it's overtaking us, but we still don't seem to be properly aware of it. The European Union, China, the United States, India ... there are plans, now almost urgent, that should have been implemented at least two decades ago, yet only now are we starting to move in a more or less serious way. More 'less' than 'more' ....

Those of us who work on the problem (in my case, on the impact on marine fauna and flora) have now understood that at this point the effects of greenhouse gases are irreversible in the medium and long term (always from our human "term" view), and that we must try to change the discourse not only by mitigating but by preparing everyone for adaptation. The inertia of warming in places like Antarctica will be very strong, and even if emissions are reduced the whole system will follow a positive feedback loop that is difficult to predict.

Difficult, yes, but not impossible. And the models are becoming more and more specific about the error ranges, thanks to data from the present, narrowing the past and the future, what lies ahead. The biggest problem is controlling the velocities. We don't know how fast changes will occur, even though those changes are already taking place. What we do have to be clear about is that we probably do not have the degree of capacity to adapt to such changes that we would like. That is to say, many of the things that are already changing will do so at a speed that will not allow us to assimilate the transformations in time, which will lead us to a series of collapses that will undoubtedly reduce our quality of life (not our existence, but a model of doing things to which we have become accustomed).

The most curious thing is that in the countries where the decisions of our destiny are made are still people who do not believe in climate change and the global problems that are beginning to overwhelm us. In the United States, in 2010 only 59% of people believed in the existence of climate change. In 2017 that figure had increased to 73%. As if it were a fad; as if it were something we should be interested in for a while but then tire of. Among the more conservative (such as Donald Trump's supporters), the figure drops to 23%. These are people with considerable global and individual power, people who need to be made aware that things are happening, that it's not a conspiracy of hangers-on who simply want a 'happy' world full of good intentions and little flowers. It is a threat that more than 99% of scientists from all over the planet devote their efforts to transmit in all its gravity, because they already know that it is happening (Fig. 12.1).

Just a few years ago there were highly skeptical groups, which have now retracted with the excuse that "climate change does exist but not of anthropogenic origin". Either they are clueless or a bunch of morons: I am inclined to think the latter. I agree about critical thinking, that as scientists we should be skeptical and that there are many things that need to be resolved because we do not understand or cannot explain them. But it reminds me of all those scientists who initially refused to blame tobacco for various cancer-related

Fig. 12.1 Swimming pool on an iceberg

ailments. For decades they tried to create palpable confusion about cause and effect until it was impossible to deny the obvious. In this case, the tobacco industry had a real interest in delaying by decades the inevitable: that society should be able to exercise judgment over tobacco consumption and have the power to create a state of opinion necessary to improve the health of those who did not want to smoke passively or who wanted to be informed of their options if they overindulged in consumption. The same happened with leaded gasoline, with certain additives in food, and with insecticides ... a group of scientists was gathered, paid and put to write ambiguous articles that were not very replicable and often meaningless.

But science gets there—it always gets there in the end. Science doesn't care what you believe ... it's like that, for better, for worse. One day or another, it arrives at the exact result. In a certain sense, something similar happens with climate change, with a few who through their skepticism delay the achievement of a firm answer because of economic, social and even anthropological interests. They delay any decision with debate that today is already proving sterile, because we need solutions on a global scale right now. They gain time for the companies that pay them, time that we do not have.

Where does Antarctica fit into this complex equation? Almost in the center, even though it is more than 12,000 km from Europe. Antarctica's profound sense of isolation is false. When you are there (as I will describe later), you feel that you are at the end of the world and find it hard to believe that what happens on the white continent can affect us. But it does. Everything points to widespread melting, an erosion of the ice masses that is accelerating, as for instance the Thwaites glacier. Stronger winds may be pulling warmer water into those huge ice masses, which in turn leads to increased rainfall (rather than snow production), which leads to more melting.

We are not talking about radical changes yet, but sometimes not very noticeable increases in a few tenths of a degree that lead down a path of no return. And they are not symmetrical. For example, Minji Leo of the Korean Pukyong National University explains that the albedo (the proportion of sunlight reflected by ice) is decreasing between -0.0007 and -0.0015 units per year in West Antarctica, yet in East Antarctica it is slightly increasing by +0.0006 units per year. In some areas, as we have already mentioned above, the melting is inexorable and is accelerating, while in others it seems to be slowing down and the ice mass even to be growing. Let us not forget that it is a continent and so the amount of ice is really huge, thus spatial heterogeneity is thus guaranteed.

This melting of ice causes changes in currents in the areas where it occurs. A team of Australian scientists detected a drastic reduction in the amount of deep Antarctic water on the coasts of the white continent in 2010. This is where the cold water is formed that rises in other places on the planet, forming a silent ribbon that connects to the entire globe and circulates for hundreds of years to end up in the same place again. The volume of this cold water has been reduced by almost 60% since 1970. Its creation is basic to the functioning of the climate of the entire planet and, because it is less dense due to melting ice, it cannot sink and flow toward more northerly latitudes in the same way as it did until now.

Consider that more than 90% of the energy stored on our planet (outside the Earth's crust and lower layers, i.e. the 'living planet') is stored in the water of the oceans. Climate change is, in fact, an oceanic change, the Antarctic being an essential place of CO_2 absorption and one of the main sources of cooling of the planet. Although specialists do not yet fully understand how climate regulation is brought about by the ocean masses (there being two hypotheses, that of thermohaline circulation and changes in ocean–atmosphere dynamics in the tropics), it is clear that the melting of (continental) ice influences the entire planetary system.

The increase in CO_2 has certainly been linked to a decrease in ice and an increase in temperatures. Greenland could lose its entire ice mass in about 3000 years. What about Antarctica—which part will be most affected and how will it impact us? Let's keep one figure in mind: the 35 m that sea level rose on average across the planet in the Pliocene; of those 35 m, it seems that 19 m came from Antarctica. "If we think about sea level rise today," says Eric Leuliette of the United States National Oceanic and Atmospheric Administration (NOAA), "of the 3.3 mm annual rise, Antarctica is responsible for about 0.78 mm per year and Greenland for about 0.33 mm per year." In other words, one-third comes from the melting of the poles, the rest mainly from the expansion of the sea by heat absorption. However, when the process accelerates on the white continent, this figure may increase greatly. Not to mention the Arctic where, as we shall see, the changes are becoming much more dramatic.

The Antarctic Peninsula: The Most Accelerated Zone

In spite of the galloping crisis we are experiencing, CO_2 emissions have been increasing unabated. We have not changed the trend because the greenhouse gas production that has been stopped due to industrial stagnation has been replaced by the emerging sources, which are taking over, and because the inertia is such that we are not going to stop this in a simple way. The effects of such an increase are very noticeable in the more vulnerable areas of the white continent, such as the Peninsula and the east, places where an increase in air temperature of between 4 and 5 °C in the past 60 years has been demonstrated. These increases have been asymmetric and variable, creating irregular but inexorable losses in ice cover. Between 1993/94 and 1998/99, some 54 GT of ice were lost per year, equivalent to about 0.15 mm of sea level rise from the Peninsula alone. More generally speaking, a loss of about 153 km³ of ice per year is estimated, which implies a contribution to sea level rise of about 0.4 mm per year. The average sea level rise across the planet is estimated at 3.3 mm per year, accelerating since 2010 in a really worrying way (in that amount, the average was about 1.8 mm per year).

It seems that the white continent, in global figures, contributes little compared to northern glaciers. But for how long? "Over the past few decades while the Antarctic Peninsula has been warming, other parts of the continent have been cooling," says Eric Steig of the University of Washington in Seattle, "but East Antarctica is beginning to warm at a rate as alarming, if not more so, than the Peninsula itself." In this area the temperature rose by only 0.1 °C per decade since 1957, but in the last 10 years this increase has begun to grow exponentially. The ice cover is not only retreating but is causing that positive feedback loop that we were talking about, which causes a mismatch in the balance between snowfall and ice melt. Accelerated melting could undermine the firm anchors of the immense ice masses far from the coast, which could then collapse. Inland lakes are created in the glaciers, sometimes visible from the surface, sometimes widening inside, invisible from the air, further accelerating melting. An area previously thought to be very stable seems to be less so, and about 10 km³ a year may be added to what has already been lost. Warmer seawater is reaching the glaciers and is relatively easily breaking up an area already accelerated by atmospheric influence, where temperatures are rising steadily. This slightly warmer water is also causing more humidity, leading to more precipitation. "Also recurring phenomena like El Niño will influence higher precipitation," argues Jens Grieger of Freie Universität Berlin. "In long time series we detect increased rainfall in areas like the Peninsula or West Antarctica."

The models speak of the melting of Greenland around 3000 AD, if there is no change in the trend before, with a theoretical increase of about 10–15 m in height. With 550 ppm of CO_2 (unthinkable; but I remind everyone that we are already above 415 ppm), in 2500 AD the average temperature of the planet will have risen about 3.8 °C, and if we reach 1000 ppm of CO_2 (a huge catastrophe, according to our present forecast and that has been recorded in other eras, such as in the Eocene), the average rise would be about 7 °C. It is an average temperature, which implies that there are peaks and troughs … The rise in sea level will then be between 0.5 and 1.2 m greater—and, I repeat, this is taking only Greenland into account. What if part of Antarctica starts to melt as well? Let's multiply that figure by five or seven (nobody puts the total meltdown of the immense frozen mass of the white continent into the models, although it seems that this has happened before). In fact, if this positive feedback effect continues, sea level across the planet could rise to about 25–35 m in about a thousand years.

Sea level rise is a highly complex issue, subject to continual calculation and re-calculation. At present, no one has the magic number. It remains to be found whether the measurements made by satellites, glacier monitoring and other technical means are entirely accurate, especially because it is not just a matter of tracking how some ice melts. It is about a series of actions and reactions of which the nature and consequences, the speed and the exact mechanisms are not yet understood. Meanwhile, the steps toward change are accelerating and we lack the time, the means and the coordination between scientific groups to understand where those 3.3 mm per year of sea level rise, now a reality, are heading.

Poles Apart

Why are there asymmetries between the North and South Pole? What is the reason for the Arctic seeing such drastic changes in a short time and in the Antarctic taking longer? In the past, warming at the poles has not been simultaneous either, and paleoclimatic models have shown that the changes (accelerated in the north) end up having repercussions in the south. The connections are atmospheric and oceanic. Everything is connected by these invisible ribbons that transfer heat or cold from one point of the planet to another, accelerated to varying extents by marine currents. They may take hundreds or thousands of years to occur, but at a certain point there is no return. The deep ocean connection is undoubtedly the most important function for understanding the impact of one pole on another, demonstrated in part by the existence of some 230 species at more than 14,000 km distant.

The cause of the asymmetry is varied. The Arctic has more sea; in fact, apart from the huge island of Greenland, it is all sea, capable of absorbing much more heat than a frozen continent like Antarctica. That is one of the keys to understanding the differences, but not the only one. In Antarctica the hole in the ozone layer can extend to 90% at certain times and areas, while in the Arctic it never reaches such values. Ozone has to do with the capacity to retain the cold and not to heat up the atmosphere. As we have seen, the decrease in ice cover will change energy flows. "While the Arctic has lost, on average, about 57,000 km^3 of sea ice per year from 1979 to 2017, Antarctica appears to have gained about 11,000 km^3 of that same seasonal ice per year," says Ted Maksym of the Woods Hole Oceanographic Institution (United States). And thickness—that is, multi-year ice—has also been lost. In the Arctic, 40% of this thickness was lost between 1980 and 2000, with the loss accelerating to as much as 0.6 m per year in certain areas.

Ice, especially when covered with snow, has a higher albedo (as mentioned above, the ability to "bounce" sunlight back): ice with snow has an albedo of up to 0.9, while water alone has an albedo of no more than 0.05. Less ice cover means more warming of the water, and thus of the whole system. This creates positive feedback: less ice means more heat absorption, which melts more ice, which causes more heat absorption. But as we have seen in the ice, algae can congregate and create an ecosystem of their own.

Let's focus on the Arctic and its changes. In a single liter of ice there can be up to 1 mg of chlorophyll concentrated in a few centimeters of depth, which is very high compared to the 0.005 to 0.010 mg per liter in the rich polar waters around it. Many organisms that depend on this food source are affected, such as krill. In this case, this tiny crustacean draws not only energy from these algae but sun protection, its resistance to ultraviolet rays. Therefore, changing the dynamics in these polar ecosystems would put it in check. "The sea may become more dependent on a planktonic food web with different species, and more similar to what we have today in lower latitude seas. The circulation patterns of both water and air will probably also change, with the consequent changes in climate and energy flows in the different ecosystems affected," says Carlos Pedrós-Alió of the Institute of Marine Sciences (CSIC).

On land, in the Arctic zone, changes in the cryosphere are also noticeable. Permafrost, the permanently frozen layer of the northernmost and southern-most latitudes of the planet, covers very large areas, especially in Siberia and Canada. In many places, an increase in temperatures and the thawing of the surface areas have been detected for some time. "This has repercussions in the acceleration of organic matter degradation processes, a change in the carbon cycles in these soils thanks to bacterial action, which would put into circula-tion, among other things, more CO_2 by respiration, contributing even more to the increase in temperatures," says Richard Spinard of the National Oceanic and Atmospheric Administration (NOAA). It is therefore the changes in the 'small' that are the focus of scientists' attention, because that is the basis on which bears, seals, caribou and the humans who inhabit this area of the planet are sustained.

The chief disruption will undoubtedly be to the ecosystems, on several facets: (1) with much of the ice gone, the phytoplankton cycle that depends on that ice would also disappear, altering both the amount of food available and the food chains; (2) Many animals such as the polar bear would be forced to move and change their habits due to an absence of ice platforms for their subsistence; (3) The forest mass (doubled in the last two decades) would become drier, increasing the risk of large and uncontrolled fires (as in Canada and Russia, for example); (4) Many cultures whose way of life is based on coexistence with ice may disappear or be forced to change habits; (5) The Gulf Stream could be altered by the change in water temperature, in a process of slowing down that is already influencing the general climate of the entire Atlantic area of influence.

Let's focus on this last point for a moment. How does such a phenomenon occur, and what are the consequences? The story goes as follows: if the ice of the glaciers (especially Greenland) were to melt, a layer of cold, fresh water would be created, and this would partly prevent from rising in the Gulf of Mexico (where it warms) the Gulf Stream, a current that distributes its heat to northern America and Europe. The lack of heat transport would cause temperatures to drop, which could feed back to create the opposite effect of what we see now: a new glaciation.

This is how scientists like Rainer Zahn of the Institute of Environmental Science and Technology at the Autonomous University of Barcelona in Spain believe the system works. At a certain point, the thermodynamic machinery goes from overheating to drastic cooling. The worst thing is that these changes have been seen over just a few decades. We still need to understand how and at what real speed, but it is clear that we are already beginning to know that the planet alternates between warm and cold periods. I do not wish to imagine what would happen if the process were really reversed in a 100 or 200 years. Worse than a torrid world, we would be covered again by hundreds of meters of ice advancing non-stop across the northern hemisphere.

Human beings are already looking at the 'positive' side of the diminishing sea ice in the Arctic and certain places in Antarctica. In spite of the fact that there are and will continue to be many technical difficulties in the exploitation of mineral and energy resources in the Arctic, from Russia to the Americas, including Norway and Canada, countries have already made symbolic gestures of their presence, where some scientists and technicians (not all of them agree on this fact) claim up to 25% of the planet's oil reserves. In spite of having much clearer legislation than in the Arctic, movements are also being detected in Antarctica. The British Foreign Office has already formally requested sovereignty over more than a million square kilometers of the frozen continent, which has particularly annoyed the Chilean and Argentine governments, whose areas of influence partially coincide with those of the United Kingdom. Shipping companies see the advantages of the thaw in the much shorter term, and already the northwest route is routinely crossed without the use of an icebreaker for a short period of time, and a similar situation is expected shortly on the Siberian route.

The point of no return for this situation, on a human scale, will be when the dynamics of melting ice changes the global cycle of the ice itself. At this point, it will take millennia or hundreds of thousands of years to return to a situation similar to that of just a few decades ago.

13

What Happened in Larsen?

A Place on the Peninsula

In January 1995, 4200 km² of ice collapsed on the Larsen Shelf in the eastern Antarctic Peninsula. In a few days, the ice fragmentation became total, satellite images witnessing the disappearance of the area as it was known until then. In the summer of 2002, a new collapse fragmented and dispersed the ice, leaving a free zone of more than 12,500 km². Nothing will ever be the same again in this area of the planet where, in less than a decade, 'permanent' ice has vanished in an area similar to that of the island of Menorca. The collapse was surprising in its magnitude but not for its origins. For decades the air temperature in this area had been rising slightly above 0 °C. "The average temperature at Larsen was about +0.2°C in 1992–93," comments Helmutt Rott of the University of Innsbruck (Austria), "but in 1994–95 temperatures increased, on average, to +0.6°C."

Significant, isn't it? For there to be a balance between what is gained and what is lost, one of the main factors in these huge icy expanses is, logically, the temperature. Around 0 °C, the balance in summer is positive, even if it is a narrow balance between loss and gain. From there upwards, the water lost due to melting ice is greater than that gained through the contribution from snow. "Although not much, the fact of having sustained temperatures higher than necessary to maintain this balance," explains Rott, "was enough to create the collapse. On the surface, but especially in the interior and the most basal part of the immense icy structures, water flowed, creating the ideal situation for the ice to crack. The dynamics is stronger as you get closer to the edge, the part that is in contact with the sea, because there accumulates the effect of the

© The Author(s), under exclusive license to Springer Nature Switzerland AG 2022
S. Rossi, *A Journey in Antarctica*, Springer Praxis Books, https://doi.org/10.1007/978-3-030-89492-4_13

hundreds of 'rivers' that often run invisibly through the interior of the structure. The mass balance takes into account what is eroded in the lower part and what accumulates in liquid form. If the balance is negative, the problems begin. We went from 180 mm per year of ice accumulation in the form of frozen snow to less than 70 mm per year. Considering that erosion was already about 200–250 mm per year, it is understandable that the process was hastened.

Added to all this is the effect of tides, which manage to erode what at first glance seems invincible. These incredible ice masses are exposed to both atmospheric and oceanic phenomena much more than their apparent permanence to the human eye. Larsen is part of that 11% of the total area of the white continent occupied by the permanent shelf. Unlike the sea ice, this does cause sea level to rise if it cracks off and drifts away in the form of icebergs. The northwesterly wind blew strongly in both 1995 and 2002, adding stress to the entire structure, which eventually collapsed. These ice masses can withstand occasional high temperatures perfectly, but if the conditions of mild temperatures, high humidity and warm katabatic winds are prolonged the fragmentation is accelerated. Thousands of square kilometers of ice vanished, in some cases in less than 33 days. However, the erosion process had been going on for a long time, and the consolidation of the thinning process of the ice shelf due to a progressive increase in temperatures over the last century had been observed already in 1990 (Fig. 13.1).

The first to reach the disturbed area shortly after the 2002 collapse were the Americans. Under the leadership of Professor Eugene Domack of Hamilton College in the United States, several scientific groups were able to penetrate the area flooded by fragmented ice, which made navigation very difficult. They surveyed with underwater cameras, collected sediment cores (cylinders several meters long), analyzed the water in the transformed area and reported on the situation in an area where sunlight and microscopic algal dynamics had not reached for the past 10,000–11,000 years. "During the Holocene period (ours) the collapse of Larsen B is unprecedented," explains Domack; "both carbon 14 and the absence of diatoms and oxygen isotope values in the foraminiferal structures show this." In this area they found very little life and an abnormally high grain size in the sediments, due to the influence of glaciers and the absence of primary production (microscopic algae). The thickness of the diatom layer in the sediment hundreds of meters deep on that first expedition was almost nonexistent; the system simply had not yet opened to the life observed in areas where ice came and went with summer and winter. "With the photographs we were able to observe some fauna," continues the scientist, "but in a very scattered way and only opportunistic organisms." The satellite

Fig. 13.1 Larsen A, where the air temperature reached more than 10 °C

images also show that primary production produced hardly any algal blooms, unlike adjacent areas on the same peninsula.

After the expedition by Domack and his collaborators, scientific interest increased. Despite the fact that a sea of unknowns was opening up, only a few ships, which could be counted on the fingers of one hand, could access Larsen. One of them was the German oceanographic vessel *Polarstern*. Led by Professor Julian Gutt of the Alfred Wegener Institute, an expedition on this ship managed to penetrate Larsen B in the summer of 2006/07. What kind of life was developing, what were its adaptations, what effects was the disintegration of the ice producing, what were the underwater topography and currents, and what were the effects of the disintegration of the ice? Do underwater topography and currents influence the conquest of new frontiers open to organisms that lived there? How are the new open areas connected? How do warm-blooded animals adapt to this new situation? There were many questions and very little time to gather data: barely 2 months, with only a few weeks in the Larsen area.

In early 2007, the *Polarstern* crew members were able to feel the rising temperatures in the area at first hand. "On January 20, we reached +3°C with a relative humidity of 50% (very high for that part of the planet)," explains Gutt in his expedition report: "The katabatic winds were warm and exceeded 30 knots." The scientists began their data collection. Elisabet Sañé and Enrique Isla, from the Institute of Marine Sciences (CSIC), found sediments that had only a thin film of chlorophyll and lipids on their surface. As Domack had found, very little food was still available across a vast area in which primary production was struggling, in the midst of areas where ice and water intermingled, not allowing the system to function at full capacity. "In the first centimeter of the sediment we found some organic matter deposited, but at 3 cm deep, nothing," explains Isla. The pigments come from the surface of the sea, from where light now arrives with a certain regularity. The nematodes (small worms that live in the sludge) increase their presence as food reaches them. "In areas where the ice disappeared only a few years ago," Armin Rose of the Alfred Wegener Institute told me, "there is little variety and few individuals, but their advance is unstoppable." The first to arrive, as always, are the opportunists, those that grow fast but have little ability to compete. Dorte Janussen of Germany's Senckenberg Research Institute and Nature Museum has also found something interesting. "There are sponges from bathyal zones, very deep, at just 200–300 m depth." Many organisms accustomed to a life of food hardship were oppressed under that immense ice sheet, deprived of the enormous productivity of the Antarctic summer that we have been talking about throughout the chapters. But, once the seasonal ice disappears, they begin to invade the areas outside, on the border with the ice floe. "The colonization process will be long and complex, but it's clear that as the ice disappears, new opportunities for many organisms gradually open up," explains Gutt.

Removal of millions of tonnes of ice must have an effect on both surface flora and benthic fauna, which have relied on a rain of food from above. "You have to take into account that not only the production of microscopic algae increases, but the disturbance by icebergs and the orientation of the main currents," adds Isla. In 2017, another iceberg broke off. More than 150 km long, A68 broke off from Larsen C. It was supposed to hold for a few more years, but a 200-km crack gave way. The dramatic changes are astounding scientists, who see how a direct effect of climate change is providing new spaces of conquest for many organisms.

From the trawl fisheries and the videos taken from oceanographic vessels, a picture can be seen of a fauna that is still in transition. "Many organisms in this area are largely considered to belong rather to abyssal zones," confirms Gutt in an extensive reference article, "i.e. organisms accustomed to oligotrophic dynamics, practically devoid of food." Sponges are mixed with other abyssal and pioneer-type organisms in a sometimes shocking mixture of two worlds meeting and one disappearing. "To understand the changes that are going to occur," Gutt continues, "you have to understand how long they were isolated by the ice that covered them, how far they are from the populations at the edge of the icy margin, and how much primary production (how much algae) is needed to be able to feed the system."

The changes are more rapid than in other areas. Perhaps because the door was opened suddenly, the collapse of the ice has been very fast. Only a few very restricted spots seem to be resisting the onslaught of newcomers. Whales, penguins and seals have begun to arrive as well, attracted by a functioning system where fish and krill are beginning to abound. "For them this direct and palpable effect of climate change is beneficial: you go from a near-desert, nutrient-poor, light-starved system with gentle currents to an area of much more accelerated dynamics and increasing richness," observes Gutt. At the edge of Larsen B, hundreds of taxa have been waiting for this moment to expand their frontiers. Not all will arrive at the same time or have the same opportunities. Mobile organisms such as fish and crustaceans will be first, and sponges and corals will arrive much later because they have a much more moderate rate of dispersal. "Increased production will create the conditions to replace life forms more like those we have in the deep sea with a more 'Antarctic' fauna," concludes Gutt, "but the increased ice in the area and the influence of nearby glaciers and the large amount of mud they carry may make this task more difficult if the collapsed area spreads too far or the ice dissolves too quickly." Life will be creating its own path of conquest in an area that has gone from total gloom to exposure to a whirlwind of organic matter, ready to inject new species into the Larsen Zone.

On the Peninsula, it is estimated that 18% of Antarctic sea ice has been lost in the last 50 years. Iceberg A68 alone represents 10% of the total area of Larsen C. And it has broken off early. "Too early, perhaps," says Mathias van Caspel of the Alfred Wegener Institute. "Larsen also has consequences for the cooling of water masses and the disappearance of this ice and the increase in temperature could be favoring changes at the global level." With an average atmospheric temperature increase of 1.5 °C over the entire planet, the days in which the temperature is above 0 °C on the Peninsula could exceed 130. What was supposed to be happening in a century's time at Larsen C is happening in 10 or 12 years. This is too accelerated to permit any moderately reliable forecasts. Nature is taking its course, adapting to the huge amount energy that we are injecting into the system.

A Unique Expedition (R/V *Polarstern* ANT XXVII-3)

The end of the world exists. It is far away and, even with our modern means, it is almost inaccessible. In Antarctica, reaching Larsen C (beyond Larsen A and B) by ship is almost impossible. In fact, when we arrived at 66° 12.73′ South and 60° 17.00′ West, those of us aboard the oceanographic vessel *Polarstern* realized that no one had ever beheld the scene before us. The hundred or so pairs of eyes that kept peering out during the first days of March 2011 were aware of the uniqueness of the moment. During those days, I spent long stretches of time outdoors (I'd be lying if I said hours … it was too cold) to see the ice, sea and land that made up one of the most beautiful landscapes I've ever seen. The destination that was reached unexpected; it had not been planned to go this far. The *Polarstern* had been content to return to Larsen A and B, areas already explored in 2006/07. The impenetrability of the Larsen area had to be taken into account. "Since 2002 the Americans have not been able to revisit this interesting and inaccessible place," says Rainer Knust, the leader of expedition ANT XXVII-3; "it is an achievement." It was, even more so because of the amount of work that could be done, the new findings and the chance to continue with avenues of research that had already begun to bear fruit (as mentioned above). The tacit race between the United States and Germany was being won by Germany. It was not in vain that they had insisted that they had in their hands data, samples and analyses that the Americans had not been able to repeat.

But getting so far down, reaching the remote, came at a price. The wind began to push furiously on the ice, which prevented us from reaching the innermost point of the inner part of Larsen B, which we all wanted to reach to check the evolution of the physical, chemical and biological conditions of the area after almost 5 years of absence. The expedition manager gathered us together and somewhat hastily laid out the problem. The expedition meteorologist, Michael Knobelsdorf, had painted an almost apocalyptic picture: "If we don't get close to the tongue of ice and land that divides the area in two, we could be trapped for weeks by the ice that moves due to the strong winds from the south," he explained. Knust, much calmer than his colleague, added that it might be too risky to reach the Larsen B station. Somewhat dismayed, we gave priority to not becoming stuck in the middle of the ice, which would have held up the expedition and been a worry for people.

Many of us wondered how the weather had changed so quickly. Just a week and a half ago, we were enjoying abnormally good weather in Larsen A. In our shirt-sleeves, we were watching climate change live: temperatures of up to 8–10 °C melted the ice, forming spectacular waterfalls in areas where the ice reached the edge of the sea in cliffs tens of meters high. The splendid sun accompanied that strange, unreal feeling of being in the right place but in completely inappropriate conditions. To maintain the equilibrium of the ice masses, tolerable temperatures are not much above 0 °C, as explained before. Reaching 12 °C was something that even those of us who are not used to Antarctica (there are people far more experienced than me, I assure you) perceived as anomalous. I was thinking for a long time not so much about the consequences (as I have been saying, nature progresses not so much by very punctual changes but by long trends) as by all those who deny the obvious or, in another equally imbecile or self-serving strategy, who minimize it. The marvelous sensation (from a visual point of view) was mixed with a disturbing one (from the point of view of perception of change, of collapse). The winds were blowing from the north, dragging clouds down the mountains to form a magical landscape. But the wind was hot, compared to what it should have been.

It spurred us on. For all of us who understood what we were experiencing live, the fact of being there, seeing what we saw and feeling it made us work even harder, to take more samples and to try to understand as quickly as possible why the Larsen area is changing and what it could mean for the future of this area of the planet and others like it. We observed the changes, we saw the progression of organisms underwater and we took sediment samples again. To be able to make the pertinent analyses of the matter available for the organisms, we studied the succession stages of the organisms and were able to fish from the bottom of the Larsen C sea for the first time (Fig. 13.2), where the

Fig. 13.2 A rare photo of Larsen C, one of the most unreachable zones on Earth

diversity and abundance of suspensivores (sponges, corals, gorgonians, etc.), crustaceans and fish was at a minimum. In this remote and still uncolonized area, seals and penguins had not yet arrived, let alone any whales. More territory to conquer, new horizons where the new colonizers, when environmental conditions permit, will begin to transform bottoms cleared of the ice cleared that had so long made the area a place inhospitable to most species.

14

Organisms and Climate Change

Changes at the Base of the Chain

When we arrived at the Jubany base at the northern tip of the Antarctic Peninsula in March 2011, I saw with my own eyes that something was different from 11 years ago. On that expedition in 2000, there were no visible land plants. The ground was basaltic rock, giving a black background dotted with white snow and ice. In 2011, there was a meadow. The ground was green. Not surprisingly, I wasn't the only one to notice. People much older than me who had been going to that base since the mid-1980s corroborated that there had been a transformation to the landscape of the Peninsula and its adjacent islands.

In part, the presence of plants was due to new species, invasive ones (as we will see a little later), but that was not enough to explain the conquest of space in a land so accustomed to the cold and the added imposition of a highly seasonal environment. The sustained increase from 4 to 6 °C in the atmosphere over 50 years was vividly reflected by the presence of organisms that previously had been unthinkable, and now they were at thriving. But, as they say, this was basically anecdotal, and just the 'tip of the iceberg'. "More intense rainfall, more constant winds, more snowmelt … and a higher temperature: the perfect recipe for terrestrial plants to feel more at home," explains Matthew Amesbury of the University of Exeter. "If we look at biological activity over the last 150 years, we see a considerable jump along the Antarctic Peninsula, especially from the 1970s onwards." In this sense, the native flora has found more favorable conditions to grow and is steadily doing so wherever it finds a niche. "The native plants are very accustomed to the harsh environment," says

© The Author(s), under exclusive license to Springer Nature Switzerland AG 2022
S. Rossi, *A Journey in Antarctica*, Springer Praxis Books,
https://doi.org/10.1007/978-3-030-89492-4_14

Dr. Znçoj of the Department of Antarctic Biology in Poland, "but if conditions continue to 'soften', we may find more invasive plants thriving in the medium term."

It is in the sea, however, where possibly more changes are taking place, because outside (for the moment) there is hardly any space for colonization due to the extensive ice cover that continues to act as an obstacle to the invasion of more remote areas of Antarctica. As we commented earlier, the waters of the Southern Ocean have been warming and becoming fresher due to continuous melting of ice from the most vulnerable areas, the northern ones. In places like South Georgia the water has warmed by 2.3 °C over the last eighty years in the topmost 150 m, with the change being more pronounced in winter than in summer. Concentrations of phytoplankton, the microscopic algae, have unexpectedly decreased on the west coast of the Peninsula yet have increased in more southerly areas. In both areas the ice cover has decreased, and in the northernmost part of the Peninsula the ice cover has disappeared locally. "These are asymmetrical situations," observes Sébastien Moreau of the Université Catholique de Louvain; "in offshore waters we have gone from 0.73 to 1.03 GT of carbon per year, but these calculations vary greatly from one place to another.

As we have discussed before, environmental factors are important, and it seems that light is one of the main players in biomass generation. Scientists such as Hannah Joy-Warren of Stanford University have long studied this, weighing the effects of temperature, nutrient availability and incident light, as a few examples. "There will be algae that adapt, there will be algae that adapt less, the important thing is to understand, at the end of the day, whether the matter they produce will be less or more for the next trophic levels."

In the water column, plankton organisms in the upper layers respond much faster than life in the benthos, and accelerated changes are being detected in many places simultaneously. Those algae whose cycle is entirely dependent on ice are beginning to disappear, with a cascading effect affecting all organisms in the food chain, from krill to seals. *Pleurogramma antarcticum*, the silver anchovy-like fish on which so many organisms depend, has been disappearing from several locations, displaced by the new environmental conditions that are not as favorable as in the north (Fig. 14.1).

Mark Moline of California Polytechnic State University in the United States reflects this change in his long time series, spending years tracking phytoplankton and other organisms at bases in that part of the world. "Diatoms tend to be scarcer than other algae, the cryptophytes, which do not depend so much on the dynamics of the ice," explains the specialist. It must be understood that this is a change in size and therefore in the availability of food for

Fig. 14.1 The icebreaker *Polarstern* entering the ice sheet at Austasen

other organisms in the food chain, both through the manipulation of that food (as we have already seen) and by the amount of carbon ingested. Cryptophytic algae are not preyed upon as efficiently by krill, which leads to a greater abundance of other filter feeders such as salps, which seem not to be hindered by the larger size when ingesting algae. Salps are less nutritious than krill—they contain less matter for fish, penguins or seals, reflected in their energy balance as a decrease in their likelihood of accumulating fat or transferring food to their young.

Above certain temperatures, cryptophyte algae are more comfortable than diatoms, and they displace them in the ecosystem as they are able to survive better: salinity, lowered by the thaw, also gives them an advantage over their competitors, so in coastal systems the change has already been occurring for some time.

Some specialists, however, also emphasize what is gained. Even if there are alterations to the productivity, there are now areas with algae in which it previously could not grow because it was covered by those 2 or 3 m of opaque ice and water. "Some 24,000 km² have been gained in recent decades in the eastern part of the Peninsula," says Lloyd Peck of the British Antarctic Survey, "which has allowed the formation of blooms that have generated some 500,000 tonnes of carbon." That production has allowed the creation of another 410,000 tonnes of carbon in the form of secondary production (zooplankton), which increases the amount of biomass in the system.

So where does that leave us? Is carbon being produced or lost? If only it were that simple! What is happening is a substitution of one system for another. As always, some lose while others gain, and that is what we should be concerned about.

Polar organisms as we know them depend on ice dynamics and strong seasonality, with all that this entails in terms of the physical and chemical characteristics of the system. What we see as small changes in temperature, salinity or pH (see below) can cause a very important regional or global change. Those organisms that depend on ice and its seasonal properties will tend to disappear from many places, taking refuge in areas where ice cover changes are smaller. The Antarctic system is known to be 'bottom-up', meaning that all changes depend on the dynamics of the smaller ones, those at the bottom of the food chain. Predators, the herbivores, have only partial control over this production. Each region, each place has its own characteristics that will make individual species suffer or flourish due to changes in temperature, ice cover, and so on, but in general increasingly isolated pockets will be created in which organisms reliant on the dynamics of the ice will survive, while those who depend less on it will thrive in the areas around them. We have already seen that in areas where there is ice the algae that are most successful are those with 'antifreeze', but this type of molecule is useless if it is unnecessary to withstand extreme cold. However, there are many varieties of phytoplankton in Antarctica, so there will always be some species ready to embrace change and take advantage of opportunities.

Ice and its future dynamics, however, are not the only problem facing the organisms living on this forgotten continent. In the long term it is the temperature itself and its increase that will be responsible for the great transformation in the column, especially in the remote bottoms where other living beings are already suffering from these changes.

Benthos, Fish and Stenothermia

There is a tense atmosphere in the meeting. Scientists are talking about experiments on adaptation and tolerance to the increases in temperature already occurring on the Antarctic seabed or, according to predictions, will take place, and how they should be approached. Sometimes things are done very hastily, poorly thought out, poorly done. These results, if they do not then pass through the scientific sieves that we ourselves impose on experimentation, give rise to misinterpretations, erroneous data, subjectivities, implausible situations and bad predictions. That is why there is a tension, because we all know that physiological experiments involve a high degree of artifice yet are necessary to understand the future distribution and survival of species, what ranges they tolerate and how they face up to the pressure of change. But let's not go overboard.

One group working on crustaceans seems not to have not done its homework in approaching and executing the experiment. Colleagues are jumping all over the group's members. In general, the relevance of quite a few experiments is questioned because of their poor fit with reality. In some experiments that have been carried out, the temperature in an aquarium was increased at a rate between one hundred times and ten thousand times more than predicted by predictive models of climate change. Organisms are put into thermal shock. In reality, what we are trying to see with these experiments is the limit of tolerance—in which waters they could live and in which situations they could not—in order to calculate possible migrations or situations of isolation. It is not about their adaptation, which in many cases is impossible due to the accelerated change, but their potential migration or how a change in the thermal conditions of the environment influences their metabolism.

Most animals living in Antarctica are stenothermic. This means that they have a range of tolerance to low temperature changes; that is, they are used to very low temperatures yet, if these are changed, their metabolic system does not hold up and they die. "Temperature and dissolved oxygen go hand in hand: the higher the temperature, the more activity, but also the less dissolved oxygen in the water, which is a double problem," explains Hans-Otto Portner of the Alfred Wegener Institute. "The organisms tolerate stressful situations; the problem is when these situations become chronic," he concludes. Most of these Antarctic organisms cannot exceed 10 °C and many die at even 5 °C. It is clear that, in the long term, there will be what is called an adaptive compensation, where the change will gradually leave the more tolerant ones alive and displace the less tolerant to other latitudes.

This is not unique to Antarctica or the Arctic. Many tropical organisms that suffer in intense heatwaves move to cooler areas, and those in more temperate latitudes seek even cooler areas to withstand the heat. At the bottom of the Antarctic, temperature increases of between 0.5 and 0.75 °C are being predicted over the next few decades, which may not sound like much but will greatly affect a large number of organisms unable to withstand such a rise. "One of the problems is the energy that fish, for example, have to invest to adapt to a constantly higher temperature," says Tina Sandersfeld of the Alfred Wegener Institute. She and her colleagues have found that fish could sometimes live but that they invested far more energy in mere survival, in breathing. Organisms living in Antarctica are adapted to high thermal stability, which means little or very little physiological flexibility in the face of such change. Temperature increases, however small, will increase their respiration because they need to compensate for the metabolic acceleration.

In fact, it is not only the temperature; many factors change all at once, putting the organisms under stress: ice cover, ultraviolet (UV) rays, productivity, disturbances due to icebergs, acidification (as we will see in the next section) … enduring all these together is a huge challenge for living beings used to stability. Many of the reserves of these organisms have will have to be spent in alleviating the stress suffered by the changes, and this will have repercussions on reproduction.

We are sometimes obsessed by studying adults and the consequences of these changes on their survival, but it is just as important, if not more so, to study future generations. Larvae, eggs, juveniles, new recruits—how do they cope with change? The spaces to be colonized are altering rapidly, and organisms that until recently could reach South Georgia cannot today do so because the temperatures there have risen, forming an invisible thermal barrier. But beware, organisms that until now have been prevented from entering by the extreme cold, such as large crabs, sharks and large fish, may soon be able to enter and wipe out the Antarctic bottom stocks. Endemic fauna could be endangered by biological invasions due to temperature changes. The so-called durophagous species—that is, those accustomed to eating shells, carapaces and other protective systems—are currently not present in Antarctica, but they are already on its border, waiting. There are brachiopods, crustaceans, bivalves, which will be easy prey if these creatures become part of the fauna in this part of the planet. Over the past 40 million years they have been kept out by the low temperatures, but the rapid rise may favor their entry into the northernmost parts. This diverse world, full of forms and adaptations more typical of ancestral times, could be reduced to places like the Ross Sea or the Weddell, in its southernmost part. Water above 0 °C could be a fertile ground for them to invade from the Scotia Arc and disperse rapidly. Icefish, which form part of the diet of seals and penguins, would also be easy prey for organisms such as sharks and large teleosts. Their sluggishness, habits and slow growth could not withstand the pressure and possibly they too would suffer from local extinction.

So many changes, so little time to understand them. And I ask myself how we can reasonably study all these things in time, given such precariousness? I am sorry to say that understanding requires time, good protocols, ideas and knowledge applied to sound experiments. What I observed at the beginning of this chapter is surely due to a lack of expertise on the part of the scientists and the pressure of running a contract that lasts just 2 years, or poor funding for a project. Let's not forget this when we ask for accountability for their work from a team of scientists working at the edge of feasibility, in terms of time and money, to do things properly.

Acidification at the Poles

There is another problem related to climate change, less known and more perverse. The CO_2 that is increasing in the atmosphere is absorbed by the ocean, in part (25–30%). Among other things, the increase in this gas will lead to the acidification of both continental and marine waters (or a decrease in their alkalinity). The matter is complex, and it comes about in the following way: CO_2 entering the sea combines with water and gives rise to another compound, a weak acid (H_2CO_3), capable of easily releasing hydrogen ions (H+). Having lost this hydrogen, the carbonic acid remains as HCO-3 and, to a lesser extent, CO=3. When this happens, the hydrogen ions remain in the water and reduce its pH. In other words, the more CO_2 in the atmosphere, the more hydrogen ions will be put into circulation.

This acidification, measured as pH (or the concentration of H+ ions in the water), is already occurring in our oceans and many marine organisms will be—or already are—affected. The pH of the waters that make up our seas is slightly alkaline (pH 8–8.3); that is, it exceeds the neutral (pH 7). The downward trend in pH, particularly in certain areas of the planet and at the surface, may introduce as yet unforeseeable changes in food relationships, matter and energy flows and oceanic geochemical balances. Of all the oceans. Like that of warming, it is a global process that is difficult to reverse because acidification is a much faster process than de-acidification due to the issue of chemical equilibrium and, once it has been put into circulation, where the acid ends up.

The problem lies in the fact that the acidification we are detecting today is very rapid compared to that in other periods, such as the Eocene. In other words, pH is falling at an unprecedented rate. What is known for sure is that, relative to the so-called pre-industrial period (late 1700s), pH has dropped 0.15 points to the present day. Various observatories detect different acidifications across the globe, but it is known that the average pH decrease is around this figure.

A fall in average pH levels by 0.3–0.4 does not seem much, at first glance. But we must not forget that we are talking about logarithmic scales, since this is how pH is measured. So, what gets put into circulation is an immense amount of H+ every time we reduce the pH by 0.1 of a point. The increase in these ions is not good for the formation of calcium carbonate ($CaCO_3$), and there are many organisms that need this product to create their structures, shells, protections … think of coral reefs, for example. "We know that corals appear and disappear over the last 200 million years, and it has been found that the acidification of the oceans coincides with some of these temporary gaps in which these biobuilders disappear," says Adina Payton of Stanford University in California; "the drop in pH decreases the availability of material available to build shells, structures or protections."

The effects of acidity appear to be concentrated in marine areas where biological production is highest. At the sea surface, at between 200 and 300 m deep, acidity is much higher and more concentrated than in other deeper regions. In other words, any acidification would fully affect the most productive layers of the ocean (the three oceans surveyed indicate this), especially in the temperate and equatorial zones. It should be noted that in regions where water temperature is higher, precipitation of calcium carbonate is favored and absorption of CO_2 also increases. The balance between physical and chemical agents is not yet fully understood, but the indications point to acidification in these vulnerable areas.

Other marine regions, such as the ocean masses surrounding Antarctica, may also be greatly affected by this phenomenon since the organisms living there are already very limited in their ability to form shells. In this case, it is precisely the low water temperature that makes the 'lysocline' (the place where this mineral compound cannot precipitate) shallower: the lower the temperature, the more difficult it is to precipitate biogenic $CaCO_3$. This is a highly complex puzzle that is only slowly beginning to take shape.

Today, marine organisms have a physiological adaptation to the slightly alkaline pH of marine waters, but if acidification occurs they will suffer drastic environmental changes that many of them will not be able to withstand, the more so if it is abrupt. There are an enormous number of organisms that need $CaCO_3$ to create their structures, shells and other protections. In addition, many planktonic organisms will find it difficult to survive, such as the small algae called coccolithophorids or the abundant protozoa, foraminifera. The former are cells capable of photosynthesizing, and they constitute an important part of primary production in certain places. The latter are protozoa capable of eating algae or detritus and represent an important link in food chains. Both are at the base of deeply complex food chains, and if they cannot create their $CaCO_3$ structures (mainly aragonite and calcite) it will have consequences for their life cycle. "Also non-calcareous algae may suffer from the effects of acidification and temperature rise in Antarctica," explains Kai Xu of the University of Southern California; "some like *Phaeocystis antarctica* could suffer from a competitor like *Fragilariopsis cylindrus*." Again, what needs to be understood is what can cause such phenomena at the level of the food chain. All organisms compete for resources, so a slight change such as slightly more acidic and warmer waters can tip the balance to dominance of a hitherto cornered species.

In high-latitude waters, this acidification can be more pronounced. It is more expensive here to form the aragonite of which the skeletons are constituted because of increased CO_2 solubility, delicate acid-base balances and mixing processes due to currents. In a way, high latitudes will be the perfect laboratory to observe its effects on marine biota. In the Arctic, such changes have already been detected—not very noticeable, but palpable. Many organisms, once again, will have their defenses weakened, and those giant crabs or sharks that we were talking about earlier will find it easier to devour them.

But there are other, even less understood factors that may indirectly affect both calcium-dependent and non-calcium-dependent organisms. Acidification and other factors will also produce a change in the distribution of nutrients throughout the planet, which will influence the extent of survival of primary producers (algae) and, therefore, the circulation of matter and energy in each zone. The increase in acidification and CO_2 in the system, along with other changes, will have important repercussions on biogeochemical cycles. Early experimental attempts to understand what may happen in this respect indicate increased nitrogen fixation, thus decreasing nitrification.

This may imply less nitrogen available for certain groups of primary production but not others. Cyanobacteria seem to be at ease in this future scenario of slightly warmer and more acidic waters. The effects on the cycles of phosphorus and silica, basic nutrients for life to function, are also unclear, but it appears that they may be less affected by these changes. What will be affected is the availability of certain micronutrients. A micronutrient is an atom or molecule present in very small amounts. Atoms in their ionized forms such as iron, aluminum, arsenic and chromium (Fe, Al, As and Cr) can have their availability to living beings transformed or diminished. The decrease in OH- and CO32- can affect the solubility, adsorption, toxicity and rates of oxidation–reduction processes of metals in seawater. The solubility of metals in water is highly dependent on pH, such as that of trivalent metals, being more soluble in acidic than in basic solutions.

What is clear is that, by modifying all the trophic levels of the system, we may face unprecedented uncertainty, setting off a series of physiological adaptations of which we still know very little and that, as some groups disappear, they could be replaced by others with, for example, less energy content. Humankind is conducting a large-scale geophysical experiment, and doing it in a way that has never happened before and will never happen again in the history of the planet. The upheaval may be very important, and not everything will lose out: there are many groups of algae, such as diatoms or dinoflagellates, whose existence is not based on the formation of carbonate structures. However, the alteration due to acidification of biogeochemical cycles—that is, of carbon and nutrient exchanges in our oceans—is a real unknown. "The acidity we may reach in a few decades may be even higher than that recorded in the last ten million years," says Lord Rees, President of the Royal Academy of Sciences of Great Britain. "To go from 387 parts per million CO_2 today to over 450 parts per million in 2050—we are at 415—would already be critical," insists Rees, "but if we were to reach 550 as some forecasts predict it could mean the dissolution of tropical reefs, especially in surface waters." According to Rob Moir, Director of the Ocean Rives Institute in the United States, "Acidification cannot be a neglected issue any longer; it may be as serious or more serious than the increase in temperature or sea level due to climate change".

In a recent meta-analysis conducted by Alyce Hancock and collaborators at the University of Tasmania of experiments carried out on various organisms in the Antarctic, it was found that a concentration of 1000 ppm of CO_2 would greatly affect phytoplankton, while 1500 ppm would affect most invertebrates in these latitudes. It seems far away, but if we continue at this pace we will end up seeing it by the end of this century, or sooner. We still have a long way to go before we understand the extent of ocean acidification. It is one more problem to add to the already complicated picture that we face, and in which more and more pieces of the disturbing puzzle are coming together.

Large Consumers

I have already discussed the impact of climate on the organisms at the top of the food chain, the creatures that consume krill, fish and squid, and now I would like to point out a couple more things to complete the picture. Penguins, seals, whales and flying birds all depend on the lowest links of the food chain (phytoplankton, bacteria, etc.) for their survival, yet not all of them will be harmed. Changes to the availability of food and to the temperature itself will cause penguins such as Adélie or emperor penguins to take refuge more and more in areas of the "High Antarctic", toward the southern parts. They will follow the ice, and they will become dispersed as they will not be able to tolerate the changes. For them it will be impossible to adapt fast enough, yet other species will be able to cope with the change in a different way. For example, penguins capable of eating myctophid fish (becoming more abundant, according to the figures, because the changes do not bother them so much) are not in regression; on the contrary, their numbers are increasing. This has been proven in colonies that have been monitored for decades, and which reflect rapid changes in populations. Those not used to drastic changes will not be able to withstand the pressure, but those that are not so dependent on frost cover will stay and take advantage of the gap. "Specific diets, tolerance only to certain temperatures, breeding in localized locations … it's the perfect recipe for extinction," says NOAA's Gregory Silber; "we're trying to understand at both poles what the tolerance limits of large mammals are going to be."

Here, what needs to be understood is that some species will disappear and make way for others. Krill may be replaced by copepods or salps in many places, which in turn will influence the presence or absence of predators in those areas. Most penguin species, accustomed to a narrow temperature range and below 0 °C at the surface, are now sweltering in temperatures that reach 12 °C in any given summer (as I said earlier, I have seen it with my own eyes), something unthinkable just a couple of decades ago. The problem of the dispersion that they will experience during coming decades is the isolation of their populations, the inability to communicate and to make use of the genetic pool that allows them to subsist. Locally extinct, penguins that are unable to adapt will possibly have their role as predators taken on by other organisms (as we have seen). The same fate will befall seals and some birds that are too dependent on a certain dynamic in their lives to resist what is coming. They depend on specific feeding areas that allow them to accumulate fat, reserves for reproduction and food for their offspring. Creatures can endure bad years, but when the frequency of crises exceeds certain limits, none can cope with the extra stress and the population collapses. Factors such as pollution, fishing or tourism can be the nail in the coffin for many populations, which cannot withstand what is euphemistically called "multifactorial stress" or, in plainer words, a fragile ecosystem being screwed up from all sides.

15

The Ozone Layer and Antarctica

The Hole in the Ozone Layer

A mountain bike in the middle of nowhere. I photograph it, ecstatic, wondering what it's doing at the entrance of the former Neumayer II base in March 2000. Our guide, Ursula, head of the German base and at the same time doctor in charge of any incident that might happen during her 15 months stint, explains to me that the bike is the vehicle used to travel to the atmospheric data center, about 12 km away in the middle of the great glacier into which the base is sunk. They can't go with any mechanized means, as they would carry contamination and the system is very sensitive. I imagine the hero (man or woman) pedaling out there every day in all weather conditions to monitor methane, carbon dioxide, chlorofluorocarbons or ozone data (Ursula acknowledges that sometimes the blizzard is excessive, putting the rider's safety at risk, and then the trip is not allowed).

The data collected by these meteorological bases are fundamental to understand, among other things, atmospheric dynamics. They detect pollutants displaced from industrial zones in the United States, Russia or China (to give three examples), and they also monitor ozone. Composed of three oxygen atoms, it is unstable, but it is thanks to this molecule that there is life as we understand it on this planet. Ozone is found throughout the atmosphere, and it is in the stratosphere where it is most prevalent. It is found between 10 and 40 km above sea level, being concentrated at around 15–20 km up. Thanks to these molecules, part of the solar radiation is absorbed and transformed into heat.

© The Author(s), under exclusive license to Springer Nature Switzerland AG 2022
S. Rossi, *A Journey in Antarctica*, Springer Praxis Books,
https://doi.org/10.1007/978-3-030-89492-4_15

I am surprised to learn that our life depends on a thin layer of molecules that, in sum, is less than 3 mm thick. The problem is that it seems that we have been working hard on this compound in recent decades, reducing the thickness that we were talking about from 3 to just 1 mm in the so-called hole in the ozone layer. I say misnamed because, fortunately, not even in the most critical moments (of which there have been many) has a real hole ever been created. What is created is a zone where the concentration of the ozone molecules drops, and it can be drastic. I remember a science-fiction book in which the main characters, for fun, throw cats into areas where there are ozone holes, watching how they are burned by ultraviolet radiation. That's silly. What is not nonsense is that there is indeed a decrease in the layer's density, and that it is harming us.

In 1970 the British Antarctic Survey (BAS) detected that, in the Antarctic, the ozone data were worrying. US scientists had made a series of erroneous calculations, overestimating this presence of this gas due to faulty software that reported an error every time its concentration fell below a certain level. The problem, it was insisted, was one of instrumentation. But the BAS insisted that the concentration of the molecule above the white continent was decreasing alarmingly and that its atmospheric balloons were well calibrated. In the tropics, the ozone concentration is stable at about 250 DU (in 'Dobson units', after Dr. Gordon Dobson, among the first to calculate trace gases in the atmosphere with precise methods). By contrast, before the ozone decline the concentration was much higher in Antarctica, about 350 DU (each DU is about 10 μm, so 350 would be about 3.5 mm). At the worst points, in the southern spring of 1995 and 2006, it fell to 100 DU. Alarm bells rang, and in 1987, after understanding that part of the problem was chlorofluorocarbons (highly reactive halogen molecules capable of destroying thousands of ozone molecules), a summit was held in Montreal at which the data, experiments and conclusions were put on the table. And a consensus was reached: the use of certain types of chemical compounds had to be eliminated, otherwise the ozone layer would diminish until it became a global danger—and not only for penguins and a few krill shrimp in the Antarctic. Its depletion would influence the UV radiation reaching the planet, also the cooling of the higher layers of the atmosphere because, as I said before, the ozone is warmed as the atmosphere is impacted.

The dynamics of the ozone layer began to be understood, thanks to satellite images and the installation of dozens of points to trace the concentrations using atmospheric balloons, and it became understood that it has a marked seasonality, its lowest coinciding with the austral spring. The phenomenon did not occur at the same intensity at the North Pole. It is still not well understood why, but it is possibly due to the greater isolation of air currents in the white continent, which are trapped, making it more difficult for them to interact with other atmospheric layers.

The point is that all the studies indicated (and continue to indicate) that halogens have the ability to oxidize ozone in the troposphere and react with other chemical compounds (such as dimethyl sulfide or volatile mercury) in a highly aggressive way. Halogens are activated in part by solar radiation and, because they influence the stability of dimethyl sulfide (DMS), which is partly responsible for cloud formation, they create a positive feedback loop: more activated halogens, less DMS, less clouds, even more activated halogens, and so on. In spring, cloud cover may be present 65–70% of the time. When cloud cover is most needed, the more active halogens are interfering with the complex chemical dynamics. And if there is less ozone in the atmosphere, clouds are another real protection against UV radiation.

This UV radiation is of various types, depending on its wavelength (short). The spectrum it covers is from 280 to 320 nm. The most dangerous type for living beings is UVB radiation. If the ozone layer is too scattered, it is not able to capture the radiation that arrives, impacts on and transforms metabolic, physiological, cellular and reproductive processes (as we will see in the next section). That is why there is constant monitoring of the ozone layer. In 2008, the "hole" in the ozone layer covered a surface area of some 37 million km^2 and by 2006 it had reached 39 million km^2, an area equivalent to Canada, the United States and Mexico combined. Julian Meyer-Arnel of Germany's Aerospace Center, one of the leading experts on this type of phenomenon in recent decades, does not know how climate change will influence the equation. "At the moment it looks like we can start to see results from the restriction of chlorofluorocarbons: since 2000 it seems that the area where ozone has the lowest concentration is going down."

Jayanarayanan Kuttippurath and his colleagues at the Institute of Technology in Kharagpur, India, believe that we are on the right track: "the ozone hole is closing, the models and observations indicate that little by little, since 2000, we are substantially improving the situation of this molecule." But Martyn Chipperfield of the University of Leeds in England is more cautious: "It's true that there seems to be a trend, but we need much more time to understand where it's going," he says; "climate change is making it difficult to interpret the data and the models are not entirely conclusive in some respects."

We have a lot of understanding, research and so on ... but my question is simple. Why is it that in Montreal we managed to reach a consensus fairly quickly yet with global change we are unable to understand where we are going, even with all the evidence in our hands? Obviously, it is because the consequences of stopping CO_2 emissions, on an economic and social level, are much more profound, but it should perhaps serve as a basis for us to come to serious agreement and respect more the work of decades by scientists who have long been warning of the huge problem we already have in front of us.

The Effects on Fauna and Flora in Antarctica

What effect does ozone depletion have on Antarctica (and the rest of the planet)? It seems that, due to its isolation and the thinner atmospheric layer in this area of the world, the ozone decreases, creating a bigger "hole" in this area (note that I am reluctant to remove the quotation marks from this thinning patch). UVB rays are harmful to life. Their associated high energy causes serious damage: changes in the structure of cell membranes, in the chemical environment of the cell, in the concentration of free radicals (molecules that transform other molecules, impairing their normal functioning), in the ability to photosynthesize or breathe, or in the functioning of DNA and RNA … chaos.

As is to be expected, not all organisms respond in the same way. Some cope better, others worse and some can 'escape', while others have no choice but to put up with what is thrown at them. The resistance of organisms depends on their structure, protection, habits, capacity for regeneration and repair and, of course, habitat. But many would be harmed. In fact, it is hypothesized that, if the phenomenon goes further, there could be taxonomic changes in terrestrial and marine communities.

A 65% decrease in ozone causes up to 14 times more UV radiation to reach the planet's surface. Small changes can greatly increase the impact of radiation, harming a wide range of organisms. In Antarctica, the most vulnerable organisms are aquatic, as life in the water is more complex and trophic relationships are more diversified. It is not that seals or penguins would not be directly affected but that they would probably suffer more through the impact on algal and zooplankton communities than on their own bodies. In the sea, UV rays can penetrate more than 30 m. It depends on the turbidity of the water and the inclination of the light rays hitting its surface. In very transparent waters the penetration is maximum, but in areas where there is a high concentration of algae it sometimes does not reach a depth of 2 m. Microscopic algae are affected more in ice-free areas or where there is "fast-ice" than in those with multiannual (thicker) ice. Therefore, if algae are affected, the entire food chain is affected. It should be remembered that the maximum radiation due to the thinning of the ozone layer occurs in spring, just when the photo-synthetic production machine awakens from prolonged lethargy.

And that's not all. Zooplankton are also directly affected by UV radiation. "Both the activity and mortality of krill and small copepods are affected by UV radiation," says Steward Newman of the University of Tasmania in Australia; "activity decreases and mortality increases with increasing exposure to these high frequency waves". In this case it seems proven that the impact is direct on cell membranes, and the worst thing is that krill have no diurnal vertical migration to the bottom, unlike other zooplankton. "Krill can be at the surface during the day," observes Newman; "they don't dive to the depths every day like other relatives of theirs that make nictemeral (day–night) migrations." Krill are not alone in being affected but, again, the synergy may be decisive in view of the major changes, which we are only now beginning to understand, that are coming to the white continent.

16

Pollutants in a Pristine Place

Unintentionally Polluting?

I remember witnessing a curious trawl-fishing incident from the *Polarstern* in 2003, just before Christmas. It was late and we had already collected the samples, and we were observing some at random to check whether or not there were organisms in the sediment that might be of interest. Together with starfish, corals or sponges, quantities of mud are often collected and, in this area, we kept overtaking and being passed again by a group of scientists that wanted to emulate the passage of an iceberg by dragging a reinforced and very destructive net along the bottom, over and over again.

In that experiment (BENDEX, as we have already mentioned) they had fixed a rectangle and the ship had passed repeatedly over the same place to clear organisms from a specific area, then they returned in 2011 to see if the fauna had recovered. Well, I started looking through the magnifying binoculars and saw some curious green, shiny, hard, angular formations. In all that mud, these nodules seemed curious and unnatural. I began to manipulate them until I realized that they were glass.

Impossible. There is no glass in Antarctic sediments. I began to isolate some of these curious shapes and began mentally to connect the dots. Many passes over the same area (and therefore a prolonged presence in the same place), tiny green shards ... what on Earth—they were broken bits of Beck's beer bottles! I can assure you that there was absolutely no one else who could have left them in that inaccessible place.

It made me reflect on the immediacy of human impact on any community, especially one as remote as Austasen and Kapp Norvegia, near the German

© The Author(s), under exclusive license to Springer Nature Switzerland AG 2022
S. Rossi, *A Journey in Antarctica*, Springer Praxis Books, https://doi.org/10.1007/978-3-030-89492-4_16

base at Neumayer (which had nothing to do with the glass). Antarctica is undoubtedly the least polluted place on Earth, the least impacted by plastics, DDT, hydrocarbons or decaying organic matter. However, the mark of humans arrived both as a one-off catastrophe, such as that of the Argentine supply ship *Bahía Paraíso* (which dumped 600 m³ of fuel oil) and continuous dumping (up to 2.4 million liters per day of sewage at the McMurdo station before the treatment plant was installed). Of course, there is still a great deal of impenetrable and unpolluted continent, but the perverse effects of globalization mean that certain pollutants manage to arrive by air or ocean from industrialized places.

The anthropogenic origin of pollutants is sometimes difficult to discern, but snow and glaciers retain various pollutants that can then be put back into circulation. "You have to consider that Antarctica is rather more isolated than other areas of the planet in terms of winds and currents," explains Dr. Cripps of the British Antarctic Survey, "so the arrival of different types of molecules from the industrial North becomes more difficult than in the Arctic." Some volatile substances or even certain types of pollutants may come from major fires in southern areas, rather than from the north. But what is clear is that they arrive and settle, then remain dormant for decades.

The slow accumulation of DDT, for example, has had no effect on Antarctic wildlife so far. For many years exhaustive records have been made in certain areas to check the vulnerability of the fauna to this and other compounds. It is known that in in certain species of birds an excess of DDT can cause an increased fragility of incubated eggs and their breakage by the mother's weight before hatching. This and other side effects led to a ban on its use, at least in industrialized countries, at the end of the 1970s and beginning of the 1980s. In Antarctica, precisely because of its supposed isolation, measurements were made to check the concentration of this and other compounds. Nothing very relevant, the levels were very low, but they were accumulating year after year. Weddell, leopard or Ross seals, for example, have no more than 0.05 μg of DDT per gram of tissue, unlike the grey seal of Scotland that can have almost 10 μg: that's almost three orders of magnitude more.

Mercury, too, has reached the continent and, although in doses not as high as in the Arctic, it is considered a harmful element also. "Adult krill accumulate proportionally less methylmercury than juveniles," says Philip Sontag of Rutgers University, "which means that predators feeding on this life stage are more likely to accumulate it and pass it on to the rest of the food chain". Heavy metals can enter the microscopic algae and concentrate it in the organisms that directly or indirectly depend on them.

There is more. Cytotoxic molecules can reach places as remote as the sediments at more than 1000 m of depth. "Industrial and agricultural activity travels thousands of kilometers to reach even here, through the air or ocean currents," explains Enrique Isla of Spain's Institute of Marine Sciences, at CSIC. "We don't know very well what concentration is necessary to end up affecting organisms, but its presence is disturbing" (Fig. 16.1).

The problem with these molecules retained in the frozen part of the continent is that in certain areas they are being released all at once due to the acceleration in the melting of the ice. Heidi Geisz and her colleagues at the Virginia Institute of Marine Science in the United States have estimated that between 1 and 4 kg per year of DDT are released because of this increased melting, up to 10% in certain areas. "The levels are low, but it is clear that as a very persistent molecule it could become very harmful," Geisz explained at a meeting on polar projects. In the same working group is Rebecca Dickhut, who will coordinate efforts to look for evidence of increased contaminants in all elements of the food chain: water, microscopic algae, zooplankton, fish and penguins. Demonstrating the change and seeing how it may impact are now essential, because things are moving fast and we are still missing many pieces of the puzzle that we need to understand the picture in full. That is why Miguel Motas Guzmán and other scientists from the University of Murcia have proposed a simple but ingenious test to understand to what extent this and other

Fig. 16.1 Green iceberg in the middle of the Weddell Sea

products are accumulating in penguin blubber: a comparison of specimens from today with those preserved frozen for decades by various museums and research institutes. This will allow us to see what is happening and how it could affect the metabolism and viability of populations, especially those in the area of the Antarctic Peninsula, which are the most vulnerable.

Seals and migratory birds are also monitored. "Birds coming from other continents to Antarctica are a good indicator of differential accumulation depending on where they come from." Satie Taniguchi is a scientist at the Oceanographic Institute of the University of Sao Paolo in Brazil, and he explains that a migratory bird (an albatross or storm petrel) coming from the Australian area is not the same as one from South America. "Now we see that the accumulated pollutants do not only come from industrialized areas, they also accumulate in the Antarctic food chains, that is, directly in the southern seas." The Antarctic skua, a fearsome predator, accumulates the most precisely because it is at the top of the food chain.

A separate issue is that of microplastics. Over the decades, the millions and millions of tonnes of plastics that have ended up in the sea have been fragmented by the sun, temperature changes, biotic action and abrasion. They have become tiny pieces found in the seas of the entire planet and at all depths. In the Antarctic ice we can find 11–12 pieces of this type of microscopic material per liter. "In the Arctic we can see more than 6000 pieces per liter at times," says Dr. Kelly of the University of Tasmania, "but to find them in the eastern ice of the white continent is really worrying." Kelly's team found that chlorophyll's presence is correlated to the concentration of these tiny pieces. "Algae behave differently when microplastics are present," confirms Linn Hoffmann of the University of Otago in New Zealand; "it may be that algae produce biopolymers in the presence of these pieces, although we have yet to fully understand the mechanisms." With almost 1800 pieces of plastic per square kilometer (about 27.8 g in that area) in the waters surrounding the Antarctic Peninsula, the white continent suffers the consequences of uncontrolled dumping as a result of our unbridled consumption of this type of material. "Why are we surprised?" asks Ana Lacerda of the Universidade Federal do Rio Grande in Brazil, "90% of marine debris is plastic ...".

What is becoming increasingly clear to some people is that this vision of the continent, pure and isolated from our repeated stupidity, is less and less true. As might be expected, there is no coordinated program to track the evolution of contaminants across the continent. The effort is focused at grassroots level yet not all in the same way. This lack of coordination means that little is really known about the extent to which the white continent has been impacted by our activities. Bases are a major source of pollution, but so are tourism and cargo shipping throughout the Southern Ocean. In the Scotia Arc, islands such as South Georgia or Orkney receive a large amount of solid pollutants mainly from ships, with fishing routes being the most associated with spills. Where it is most active, tourism acts as a vector of pollution, and in some areas it is little or not at all controlled by serious restrictions and monitoring protocols (see chapter on tourism). Despite a certain degree of control, atmospheric pollutants and certain solid discharges are the order of the day.

It is sporadic, it is localized, but it is there. More worrying is the situation at the grassroots level. Despite the existence of protocols and a specific treaty, not every party follows these to the letter and others have been slow to implement them. Landscapes with abandoned stations, solid-waste dumps or untreated sewers are not uncommon in this part of the world. In many places, especially in the early days of settlement, the waste was left everywhere. Sometimes solid waste of all kinds was even left on top of winter ice so that by spring and summer, when it melted, it would be swallowed up by the largest sinkhole on the planet, the sea. Fortunately, habits are changing, and the introduction of certain materials, such as polystyrene pellets or PCBs, is now prohibited, treatment plants are being installed and the human impact is being drastically reduced.

Environmental awareness is also arriving here. What surprises me is that on some occasions it has to be forced on us by a situation that from the beginning seemed unsustainable. I insist on the sporadic, because in reality the spots that may be more polluted are insignificant compared to the immensity of the southern continent. But do we have to watch the shit (excuse me) accumulate on our doorstep before we realize how unfeasible the situation is? In a place as pristine as Antarctica (or Everest), this much filth offends our eyes. Already accustomed to generations of garbage collection, we seem immune to a landscape corrupted by pollution (visible and invisible), but for both Everest climbers and Antarctic base-dwellers the sight is traumatic.

Long Time Series at McMurdo

When Alessandra Schiavone of the University of Siena in Italy conducted the study on pollutants in various animals around the North American McMurdo station, she realized that the levels of certain compounds were exactly comparable to those of a first-world urbanized area. And it is not surprising, because the McMurdo station is actually a large town that houses more than a thousand people at the height of the austral summer. Human activity has profoundly changed the environment and, according to experts like Cripps, it may take decades to recover the communities that once roamed the bay where the base is located. The waste has origins as diverse as liquids generated from human waste, the remains of food and packaging, the transfer of fuel oil or the dumping of solids.

When the base began to be established, there were no specific plans. Or rather, there were, but no one took them very seriously. In such a small place, waste began to be a problem both because of the health and visual impact. But it was the unseen communities—those that received the continuous discharge at the bottom of the sea—that were changing. Thanks to thorough monitoring programs (in parallel to the base's own activity at the beginning), it became clear that the affected areas had only a few species, the most opportunistic and least vulnerable, living in their natural habitat. "Many organisms had become accustomed to ingesting human waste," stresses Kathleen Conlan of the Canadian Museum of Nature; "the changes in the communities were in response to the changes proposed by the engineers who were relocating the collectors, loading and unloading areas, etc."

Along with debris, large icebergs entering the bay disturb these bottom communities, which are locally primarily affected by man and ice. "Long-term monitoring has allowed us to detect these problems and bring them to the fore," continues Conlan. The introduction of a new water treatment plant in 2003 promoted changes in water quality. However, it will take decades for the water to recover to anything like before people moved to McMurdo. At that time, nobody thought about environmental monitoring or prevention measures. Of the more than 40 stations spread over 18 countries, only the United States, Argentina, Australia, Germany and Great Britain have an environmental monitoring protocol. However, the work reports are not accessible and, in the overwhelming majority of cases, their information has not been made public.

17

The Final Resort

Mining in Antarctica

In 2010 Toni Polo, a journalist and friend, published with me a futuristic novel about a little-known problem. We tried to capture humans' greed and desperation to maintain their status by the systematic environmental erosion of our environment. The idea was to have some mercenaries prospecting in the deep sea off Antarctica for a vein of tantalum. In *Cemetery of Icebergs*, a group of scientists find themselves in the crossfire between the Chinese (the emerging power) and an international mining consortium run from South Africa. Although tantalum is not found on the Antarctic continent for orogenic reasons (yes, it's difficult to pronounce), the issue of mining on the white continent is far from science fiction, and the novel makes it clear that, sooner or later, we will indeed end up assaulting this last remote and inaccessible bastion of the planet to obtain this raw material, our need for which is so pressing. We need more minerals, and the seabed or the remote Antarctic mountains could provide them.

When you get yourself seriously informed on the subject you realize that there is very little reliable information. In 2017 a book came out on the management of the white continent (*Handbook on the Politics of Antarctica*—take a look at the Bibliography at the back) in which several problems and potential solutions are exposed. You also realize how open and uncertain is the issue of the exploitation of this last continental bastion on the planet. One idea becomes very clear: nothing prevents, de facto, mining in that part of the planet. In the 1970s it was said that in a decade the necessary technology would exist to prospect and exploit certain minerals and oil resources or gas

© The Author(s), under exclusive license to Springer Nature Switzerland AG 2022
S. Rossi, *A Journey in Antarctica*, Springer Praxis Books,
https://doi.org/10.1007/978-3-030-89492-4_17

pockets in this southernmost part of the planet. A series of studies were carried out, mainly focused on iron veins and minerals such as nickel, silver, gold, copper and cobalt. The first surveys were carried out on the continental shelf for oil. A couple of decades earlier, in the 1950s, nobody gave any credibility to large-scale oil exploitation in the North Sea, but necessity pushed part of Europe toward self-sufficiency and concessions began to appear like mushrooms, especially in Great Britain and Norway. With the OPEC crisis of the early 1970s, what at first seemed too great an investment to extract an essential element for the functioning of transport, industry, urban and agricultural energy caused a large number of engineers to rack their brains to overcome the problems and ultimately extract the enormous amount of oil and gas from one of the richest continental shelves on the planet.

Could we overcome the technical difficulties in a place like Antarctica? I have no doubt about it. It must be recognized that it is much more complicated than extracting minerals from the ocean from 1500 m (already done in Papua New Guinea) or oil from 4000 m depth (already done in Brazil), and in *Cemetery of Icebergs* it is explained in detail. For me it is clear that we may be forced to do so in the near future. At the point when we realize that we are unable to find certain non-recyclable elements and that society has not managed to make the major changes necessary to allow us to survive (energy and material optimization, a radically different consumption model, renewable energies, large-scale operation of nuclear fusion energy, etc.), whatever it takes we will dive headlong into exploitation. "In the Antarctic Treaty, in the specific section on mining signed in 1991 in Madrid," explains Sean Coburn of Columbia Law School, "there is no veto or law preventing mining, only a consensus among 29 countries." The Treaty can be modified at any time by one or more of its parties. And there are countries that want to change it (Fig. 17.1).

Fortunately, the white continent does not make it easy for us to exploit it. First, less than 2% of the surface is free from large glaciers. The rest has a layer of ice that ranges between 1000 and 4000 m thick. The problem is not only the ice itself (drilling through it is complex and requires sophisticated technology), but the fact that it moves at a speed of a meter a day or more, making it impossible to maintain a stable tunnel structure. But we can blow it up. If the site is close to the coast, we can destroy a glacier just like that. Then comes the second problem, extreme cold. To my mind, this is one of the minor problems, because there are already protocols for extreme mining, such as in Alaska, Siberia and northern Canada. In the last country there are diamond mines of immense wealth connected by ice highways passable only in the middle of winter by large trucks (the routes fall apart in summer) and are

Fig. 17.1   Logistics at Jubany, the Argentinian–German Antarctic base

profitable, as are the tar sands from which oil is extracted (infringing a brutal landscape and causing ecological devastation). Therefore, we are overcoming barriers.

Antarctica is a very isolated continent, so the logistics would be complex in one sense, yet there are already places where human beings are accustomed to prolonged isolation. We are not that far away from seeing underwater mining cities from whence the last remaining non-renewable resources will be extracted on a large scale (see James Cameron's film *Abyss*). Those people will also be isolated for long periods of time. But for both the ports needed to extract and move the ore and the oil platforms located in the middle of the sea, the most serious problem is undoubtedly icebergs. In the Arctic there are also icebergs, but they are much smaller. In the Antarctic, as we have seen, apart from being larger, with climate change there will also be more of them around. An iceberg of a hundred meters is not the same as one of a hundred kilometers. There is no human force that can stop the latter, and if you are in its path there is little you can do to avoid catastrophe. But at that point there will surely be areas that are more or less protected from the currents that usually carry these giants. There are indeed obstacles, but in my opinion they are far from insurmountable.

So why isn't Antarctica being exploited? The straightforward and realistic answer, in my opinion, is that it is still too expensive. It is estimated that a barrel of oil (there would be about 45 billion barrels, according to the shallow prospecting done so far) would cost between $100 and $200, well above the price we are used to. The iron seams are not very pure, only about 35% ore, and the huge amount of coal there (an estimated 11% of the planet's total) is of low quality. But I still think this is all hot air. China has returned to coal, abundant in that country, because it is cheap, and it exploits this resource of a very poor quality. And it doesn't matter, people continue to burn it. Barrels of oil are still cheap in the artificial market that is gradually coming to an end. The required purity of the mineral depends on what we are looking for, because if it is a superconductor or a strategic mineral that is difficult to recycle there will always be people motivated purely by profit?

At the point when we run out of resources, many eyes will turn to the white continent. And it won't be just greed, in my opinion. It will be desperation. And desperation makes something that you once regarded as disgusting look attractive, provided the politicians and economists of the day know how to wrap up the bitter pill as more palatable. There are protocols, there is an Antarctic Treaty (which I talk about later) and none of this would be so linear but, as Ana Pallesen of the University of Canterbury (New Zealand) says, "the truth is that, under current legislation, it is not at all impossible to start exploiting Antarctica from a mining point of view."

In recent years we have seen a proliferation of mining concessions (some 27) at great depths in the middle of the oceans. Many countries are interested and register large areas (at the moment, 1.4 million square kilometers, more than two and a half times the area of Spain, or larger than Peru) so that they can exploit minerals despite the great technical difficulties, which are currently being resolved. In a place where we know only 0.0001% of its surface reliably (at depths of more than 500 m), we are already talking about large-scale mining. "Where there are polymetallic nodules or carbohydrates, there are often fragile animal forest systems, such as deep corals or sponge fields, pogonophores or beds full of very slow-growing bivalves," says Kathryn Miller of the University of Exeter.

Nicholas Kirkham explains why Antarctica and deep mining have many things in common. "Both areas are under international treaties, there has to be a consensus. In both places a lot of money will have to be invested to be able to do regular mining and there is also a lot of environmental uncertainty. The communities are very fragile and slow-growing. And, in addition, they are heritage sites," he concludes. Some have advocated using the Antarctic Treaty as a paradigm of what should control this deep mining.

In the end, in a sense Antarctica is a similar problem to the Moon or asteroids (except for the biological communities), but on Earth. Japanese, Chinese and Indians have sent probes to prospect the Moon as a possible source of strategic minerals. Of course, at this moment it is merely a show act; it would not be feasible to bring mineral from there to here in a profitable way. But that's our problem, it's hard for us to see things in the long term. There are elements, minerals, rare earths, especially for high technology, which are extremely difficult to recycle (some are impossible, once they have been used). It is true that in the future we may be able to do so (or substitute with other materials), but for the moment it is easier to think about where we can get them if they run out in Africa, China or South America.

Since the Madrid Protocol of 1991, any type of mining is prohibited in Antarctica. But it is an international treaty with feet of clay, not reliable from a strictly legal point of view. The CRAMRA (Convention on the Regulation of Antarctic Mineral Resources Activities) manages possible claims in this sense, but it cannot force anyone not to prospect because the territory belongs to everyone (and no one), like the open sea outside the jurisdiction of the famous 200 nautical miles. An Australian senator in 2006 asked, "Do I have to turn my face the other way while someone else is prospecting in the Antarctic? We know that there are people exploiting krill, whaling or fishing for fish that are theoretically forbidden or with protected species." Already at that moment was sensed a serious change in attitudes to the white continent.

During the past 50 years to 2007, worldwide demand for energy and strategic materials has grown by more than 50%, and it is estimated that it will grow by another 50% by 2030 at the current rate of growth. We have transformed much of the Earth, fragmented its ecosystems and alarmingly diminished its biological diversity, in part due to the extraction of non-renewable resources (oil, gas, minerals, etc.). The hope that remains for us in Antarctica is to realize that it is the only territory where we have agreed on its future. It is a place where, because of the Antarctic Treaty, the only parties that can collaborate to exploit something are the scientists: the results of their observations and experiments. As it is the last bastion of untouched nature, knowing the actual damage that mining promotes it would be a crying shame to see that, once again, we have not understood anything at all.

Pharmaceuticals from the Depths of Antarctica

A little-known type of 'mining' is related to extracting marine pharmaceuticals from organisms that live fixed to the substrate. Sponges, tunicates, bryozoans and gorgonians in many cases have a series of substances called secondary metabolites. These help them to avoid being colonized by other organisms competing for their space or being eaten by them. The substances, called natural products, have been the focus of the large school of ecological chemistry, which studies the wide variety of complex molecules that make up a universe of defenses and attacks and how these molecules help the various organisms to survive in the hostile environment that is the ecosystem, full of enemies. This school has had several aspects, some very interesting as they have allowed us to interpret many of the interactions in the animal and plant kingdoms in the complex dynamics of ecosystems. Others, in my opinion, are more debatable, and these have become a somewhat perverse exploitation of nature.

In 1951, in a scientific article Bergmann and Feeney inaugurated the potential exploitation of these natural products as elements that, in future, could be applied to medical science. They described several molecules in Caribbean organisms that could be considered as precursors of drugs with a direct or indirect application to the sector. The starting gun was fired for hopes that the sea would be a salvation for diseases such as cancer, and for infection, renal dysfunction and analgesia. The idea is excellent, but as the decades go by a real problem has become clear: the concentrations of this type of metabolites are very low, so tons of sponges or gorgonians are needed to obtain something that is only minimally profitable. The worst thing is that they can be described with great precision yet 'emulating' them (synthesizing them) is of tremendous complexity (impossible, in some cases). More than 90% of the time, the molecule cannot be created from artificial synthesis, which implies a major failure at the industrial level. By the end of the 1990s, the pharmaceutical companies were losing enthusiasm. Logically, as the cost of collecting the organisms (sometimes very scattered because they are rare or found only in certain places) is so high, the gains do not compensate for the expense.

As with precious corals (to which, among other things, I am committed), the renewal rate of most of these species is very low: they grow very slowly. There have been attempts (some well-established) at aquaculture of particular species, but the lack of knowledge of their biology and ecology has meant that most have been futile. Therefore, as in the case of precious corals or gorgonians, exploitation of this type of molecules from marine species is simply an act of mining. It is discovered, collected and plundered, making the species disappear at the local level. It is another mining mole, much less known yet more widespread than it seems.

What does this have to do with Antarctica? In 1997, James McClintock of the University of Alabama published an interesting paper that compiled what was then known about the science of chemical ecology in polar areas: very little. One of the most widely held beliefs up to that time was that bottom-dwelling organisms in such cold areas would not possess a large amount of these secondary metabolites. But upon discovering the enormous number of species that populated these bottoms (only about 25% have been described), someone began to suspect that part of the biodiversity was due to a series of highly complex interactions based on chemistry. It was extremely likely that, because there are so many different organisms, the complex molecules that helped them to survive are common among them. Moreover, not having to expend as much energy on breathing as in warmer environments, the energetic effort can be directed to grow, reproduce and, why not, to create sophisticated defense and attack molecules. The work demonstrated a large number of such molecules, and some scientists began to study them systematically.

Katrin Iken, from the Alfred Wegener Institute, describes several compounds in only a single gorgonian (*Ainigmaptilon antarcticus*) that could have developed to defend themselves from predation (they are soft organisms, covered in a thick mucus). This researcher enthusiastically reports on a series of molecules that serve both to prevent bacteria and other organisms from colonizing it (antifouling substances) and to avoid being eaten (because they are poisonous, have a bad taste or mislead the predator. At the end of the article, as in many others, in a simple sentence the author opens up the possibility that these substances may be useful for pharmaceutical exploitation. Later, other researchers go further and describe substances from no less than 290 different species collected from various parts of Antarctica, from the Antarctic Peninsula to the East Weddell Sea or remote Bouvet Island. They find tunicates particularly rich in anticancer substances, as well as several sponges (remember that they are the dominant organisms, in terms of biomass) and gorgonians.

The authors of the papers have produced more and more literature describing hundreds, thousands of molecules (to date, more than 25,000) which cannot be synthesized in laboratories yet which may act as anti-tumor or anti-inflammatory agents. The most interesting species are in the range of 250 and 500 m deep. The question I have is, how are we to exploit these substances, which so far have created only an endless series of descriptive articles that have more to do with technical protocols than pure science and interpretation? Right now, the only way that I can think of doing it is by simply collecting them. And at that depth, the only way to do so is to devastate the area by trawling.

Some time ago a company crunched the numbers on an anti-tumor agent discovered in New Zealand. It realized that to create enough product it would have to wipe out all the sponges in New Zealand. All of them. Because the concentration of the metabolite was so low that hundreds of tonnes of that species were needed to make the product profitable. It is obvious that at present that no one is going to trawl the Antarctic seabed, but what about in 20 or 30 years from now? This type of mining would be the perfect way to wipe out one of the most fascinating and complex communities on the planet faster than icebergs, which do it routinely. It may be argued that in that time it may be possible to synthesize many of the products, but I believe it would still be easier to fish them out. Few people really try to understand what they are used for in nature, so it is easier to describe them and, if possible, to patent them. I have always had faith that the sea would save us from many things, and the fact that extensive studies have been carried out on the potential virtues of certain molecules, in pursuit of a better life, has always been one of those possibilities. But it should always within the bounds of reason: either the species that produces them can be profitably cultivated or their molecules can be emulated in the laboratory at an industrial level. To extract them from the bottom by boom and bust, as the underwater mining of living organisms, is a complete atrocity.

18

Remote Fishing Grounds

Krill: Savior of Mankind?

In the 1970s, krill was thought of as one of the food sources that would save mankind from starvation, a key piece in the exploitation of marine protein that would allow us to feed a large part of humanity, like manna from the Southern Ocean. An ambitious cooperation program was set up to study its distribution, abundance, biology and how the stocks would withstand the onslaught of a fishing fleet prepared to withstand the harsh climatic conditions around the white continent. From 400 thousand to half a million tonnes a year were extracted until 1981, when the El Niño phenomenon drastically reduced catches in the area (Antarctic Peninsula). After this impasse, the stocks recovered and returned to similar catch levels (between 350 and 400 thousand tonnes per year until 1990).

But the Russians withdrew from the fishery. They saw it as unprofitable because it was hard to process the material on board or in factories on land. The shells contain a large amount of fluoride and peeling them was both difficult and unprofitable. The extraction rate was close to 100 thousand tonnes, far below the exploitation figures that the experts considered acceptable (between 2.5 and 5 million tonnes per year). The Japanese were the only ones to continue to harvest krill systematically, and 60% of the catch passed through their vessels. Since 2010, the fishery has increased slightly, with Norway and China being the main players in the area, although the South Koreans are close behind.

Fishing techniques have greatly improved, as well as the processing of these small creatures with a shell of about 6 cm, the essential basis of the

© The Author(s), under exclusive license to Springer Nature Switzerland AG 2022
S. Rossi, *A Journey in Antarctica*, Springer Praxis Books,
https://doi.org/10.1007/978-3-030-89492-4_18

functioning of the Antarctic system. The technical barriers to finding, exploiting and treating them are diminishing. And the need for a food source for aquaculture, which has grown at an unstoppable rate since the 1970s, is putting krill back under the spotlight of the renewable resource exploitation industry. Its enormous biomass could be a perfect complement to the 30% of fish that is used only to feed other fish (fish farms), equaling the exploitation of Peruvian *anchoveta* (about 7 million tonnes per year). Moreover, with the new laws on artificial colorants restricting their use for pigmentation of species such as trout or salmon, krill (which lends a pink coloration to these and other species) is seen as an object of desire by this vast industry. But the CCAMLR (Convention on the Conservation of Antarctic Marine Living Resources) recently clarified that, given the needs of the ecosystem, extraction could not go beyond 2.75 million tonnes per year (data show stocks fluctuate between 60 and 100 million tonnes around the white continent, more realistic than the billion suggested elsewhere). What are the needs of the system?

For the first time in the complex world of industrial fishing, it seems that someone is thinking about the sustainability. The bases of krill exploitation theoretically take into account the fragility of the Antarctic system's functioning, which depends on this species. First, a balance is struck between fishing and conservation; the extraction of the resource is taken into account from the ecosystem point of view (fish, seals, squid, sponges, whales...). After this approach, measures to protect krill stock are estimated so that these components of the ecosystem do not suffer the consequences of unbridled demand in an ever-expanding market (following our stupid and unnatural model of survival). Finally, changes that imply irreversible transformations during the next two or three decades are avoided.

On paper, this is all very nice. "The problem," says John Croxall of the British Antarctic Survey, "is that, even with improved extraction models, krill are far from being a homogeneously distributed animal." Croxall has studied the variability of krill, along with factors related to its reproduction, habitat type, migrations, and so on (in broad international cooperative programs), with potential exploitation in mind. "Years with low krill abundance are hard for organisms that live directly on krill, such as seals or penguins," Croxall adds. These fluctuations must therefore be taken into account not to make a fishery viable but to respond to species' biological and ecological cycles.

This is a way of applying ecosystem management to fishing, by basing the entire fishing model not on the dynamics of one species but of all those related to it—of the entire ecosystem. This approach, which is quite 'new' in the fishing industry, is not to everyone's liking because it greatly limits catches and operating models. But it is undoubtedly the future.

The problem lies in the actual respect for this type of approach. And climate change is not helping. Why? Because the accelerated rise in temperatures is opening up vast areas (especially the peninsular archipelago and the Antarctic Peninsula itself), where krill have already been exploited for some time. Even in winter ships can now reach areas that were unthinkable before. Paradoxically, the melting ice melt increases fishing, but it is also related to changes in the productivity of the system that diminish its capacity to generate biomass: less ice, less krill. The abundance of krill depends heavily on the processes of recruitment of new individuals, more than on 'predation' per se. In other words, there may be a large number of predators exerting pressure, but what really influences them is the primary production that feeds them and allows them to create new generations.

The situation could improve if large, protected areas are finally established. This is something that has been in the making for a long time but is yet to become a reality. "The Antarctic Treaty could be key to the protection of this part of the planet," says Cassandra Brooks of the University of Colorado, Boulder, in the United States, "but there are new players on the board who were previously irrelevant and are now gaining prominence. China, for example, was initially very reluctant to establish large protected areas. However, little by little, due to a real change in its diplomatic dynamics (i.e., in acting as a superpower), it seems to have understood that it has to give in and negotiate aspects in order to be able to exert real influence in the treaties."

The Chinese have complained that fishing figures were imprecise, studies were scarce and the proposed directives were based on inconclusive data. "Countries like China, Russia and South Korea are changing the rules of the game, not only because of their interest in fisheries but also because they are willing to keep a close eye on the future of the white continent in many other respects," explains Brooks. Antarctica is a not inconsiderable chessboard, and part of the future of humanity will undoubtedly be decided here. And many actors are becoming more and more involved in its management in the medium and long term.

Fishing is a focus of attention, and not a minor one. It is estimated that more than 30% of the planet's primary production (algae) ends up on our plates in the form of fish. That's a lot. And in the Antarctic, if we exploit krill, it will be even more, because we will be directly eating the organisms that feed on these microalgae. More than ever, we must use the precautionary principle with this organism that regulates the last system that remains untouched by the hand of man. This is not easy, since we need more and more protein for our own survival and well-being.

The Impoverishment of "White Gold"

The basis of commercial fishing in Antarctica follows a similar pattern to that in other places on the planet: the discovery of fishing grounds, exploitation, overloading the system, then collapse. It started with seals and elephant seals, followed by whales and then, around the 1960s, fish and krill. In the last case, the collapse has not yet happened, but it already has for the most important fish species from a fisheries point of view—and I am not talking about whales, as I dedicated a whole section to them earlier.

In 2000, on my first expedition to the Antarctic, I was most surprised to find four boats fishing around the Peninsula. I was convinced that in this remote and apparently protected place there should not be any kind of fishing. A big mistake. Certain species have been exploited for decades, such as the Antarctic hake or 'toothfish' (from a taxonomic point of view, it is nothing to do with the usual hake). It is the 'white gold' of the Southern Ocean. It lives at a depth of 400 to 1000 m and is fished with long bottom lines that can be several kilometers long and have tens of thousands of hooks, and its large size (up to two meters) and weight (over a hundred kilos) makes it an attractive catch. It began to be fished well above the Polar Front but, as stocks collapsed (in 1969 alone almost 400 thousand tonnes were caught in the South Atlantic) it began to be fished further south, approximately 1° latitude per year. There is much illegal fishing under a flag of convenience for this long-lived, very slow-growing fish whose populations around Antarctica are being depleted, one after another. No wonder, as its meat is delicious and very easy to handle, and some of you may have unknowingly bought it already. It is often used as a substitute for fish that are seldom available on the market anymore. "In Asian markets this fish is sold as sablefish, deceiving the consumer," says Xiong Xiong of the University of Pisa; "we have been able to detect it thanks to the DNA of the fish. Of course, its meat is not cheap, which is why it is called 'white gold'. It arrives frozen in the form of slices, and nobody respects its fishing quotas or the areas where it should be fished."

The worst thing is that there are the means to control the catch of Antarctic hake, but if we cannot control fishing even along our own coasts, just imagine doing so 12 thousand kilometers away. It is, once again, the 'tragedy of the commons'; that is to say, that which belongs to everyone belongs to no one and therefore its capture and its extraction are less complicated and thus far less controlled. In 2003, the CCAMLR estimated a figure of some 25 thousand tonnes of Antarctic hake being fished illegally. The high price on the market encourages illegal fishing to continue. It is known that in the Mauritius

Islands there is a well-established fleet and factory for this catch, which every year penetrates further due to the presence of ice.

The only predators that seem to really affect these fish are killer whales, which eat hundreds of kilos of the catch directly off the hook. Those same lines and boats are responsible for the death in a decade of half a million birds in the Antarctic, especially albatrosses. Fishing in this part of the planet comes at a price. In one of the most hostile seas on the planet, it is not unusual to hear that a South Korean fishing vessel of more than 600 tonnes is sinking helplessly. Five sailors perished in one accident and their companions were rescued in extremis by the New Zealand navy. The strong waves and ice, together with fierce winds, make the Antarctic Polar Ocean one of the most dangerous on the planet for fishing. However, the technology and crews are ready to continue the fishing and to expand it. "We don't know the biological patterns of the species and we are already exterminating it," complains Casandra Brooks in her personal blog. She and other specialists are calling for a different model of fishing on the white continent. "Perhaps Antarctica will be the first place on the planet where, by consensus, we will achieve sustainable fishing, based on the needs of the ecosystem and not just our own," she adds. Hopefully. Because trawling has not started in this part of the world in any systematic way. But with 97% of the world's fishing grounds exploited, overfished or agitated, how long will it take us to look to the Antarctic seabed?

19

Ecotourism and Invasive Species

A Seemingly One-Time Problem

"Tomorrow we will throw away except the absolute necessities," said Ernest Shackleton during the grueling polar expedition of Christmas 1908, a few years before the fateful expedition on the *Endurance*. How long will it be before going to Antarctica is nothing but a sightseeing tour? We have come a long way, to the point when people like Scott, von Bellingshausen, Shackleton and Amundsen would be perplexed by the influx of visitors who come to contemplate the beauties of the southernmost continent of the planet dressed as clumsy, multicolored human penguins. I wonder what we would consider essential on a ship full of people going to watch icebergs, penguins and seals for fun from the comfort of their cabin, which they walk around in shirtsleeves. A cell phone? Tablet? Let's not be cruel, though, because it is a positive that we have advanced and no longer undergo the terrible hardships experienced by our ancestors.

Still, the tourism phenomenon in Antarctica has its pros and cons. From 1965 to the present day, tourism has grown more than a little: from 58 people, to a peak of more than 42 thousand in the 2007/08 season (when there was still money and people spent it freely). It has stabilized between 35 and 50 thousand depending on the year, with the most frequent visitors to the Antarctic Peninsula area being Americans (36%), followed by Germans, English and Australians. The Chinese have been booming, reaching 11%, and tourism is linked to GDP. The influx just before the COVID-19 pandemic reached a peak of 74 thousand people visiting Antarctica for tourism (between 2019 and 2020). The average tourist is not exactly poor, and spends about

© The Author(s), under exclusive license to Springer Nature Switzerland AG 2022
S. Rossi, *A Journey in Antarctica*, Springer Praxis Books,
https://doi.org/10.1007/978-3-030-89492-4_19

12,000 € to 18,000 per person (they are not young either—65% are over 50) and prefers to travel by ship (more than 95% of trips are by this means). Statistics aside (which always leave numbers floating in our heads, sometimes uselessly if they aren't contextualized), what problem can there be in visiting Antarctica?

Let's look at the positive side before presenting the everlasting negative side. People who go to such a remote place are almost 100% environmentally friendly. That's for sure. They are people who spend money for the privilege of contemplating one of the last truly pristine places on the planet. Not only are they conscientious people, but on many occasions they will visit the bases where scientists explain interesting issues to them, pressing problems, basic science that they will later transmit to people on a terrace in Boston, Berlin, Beijing or Sydney over a few beers. I'm not being sarcastic. I find this point entirely positive. Besides, why should we have to be among the privileged few to get to this point on the planet? (Fig. 19.1)

The places visited are very few, and half of the visits are concentrated on about eight points of the northernmost part of the Peninsula, so the impact is controllable (in fact, there is tourism in an area of no more than 0.005% of the continent). Cuverville Island, the Neumayer Channel and the Lemaire Channel are some of these areas, and each year an estimated 20 thousand people pass through or disembark to see the top spots of the cold mainland.

Fig. 19.1 Bike used to visit the meteorological station in the German base of Neumayer, spotted on 2000 expedition

This is not an excessive number, and the tourists are conscientious and keep to specific places that are supposed to be controlled, so what's the problem?

Nor do I think it is fair to evaluate them by the amount of CO_2 that they emit when they travel, either by plane (25%) or ship (75%)—about 0.55 tonnes per passenger per day of carbon dioxide—because people who go to Australia or Japan also emit a great deal of CO_2, and there are far more of them. So, what is the problem?

As always, the problem is the real lack of control over the situation. The 55 thousand people arriving as tourists and crew, plus the 4 to 5 thousand scientists and technicians on the bases, may not represent uncontrolled hordes disturbing the idyllic Antarctic landscape, but they do promote unintended issues that need to be regulated much more strictly. For example, the ships and planes carry organic and inorganic chemicals, emit waste of various types and toxicities that can affect certain areas yet that, under strict regulation, can be minimized. But how can we demand strict regulation from these companies if half of the scientific bases emit pollutants in the form of fecal water, organochlorine substances and heavy metals? We are entering the eternal contradiction of 'no man's land'.

As we have seen before, it is only by the will of the people that nobody does what they should not in Antarctica because, from a legislative point of view, there is nothing firm. Many tourist companies take advantage of this and are respectful as far as their budget allows (and they are based in many countries, let's not forget). Contradictions arise, such as that of the Argentine government. After a tourist ship accident in mid-2000, it promised real restrictions on companies that profit from transporting tourists in unsafe conditions (due to inappropriate vessels and crews being unprepared for one of the roughest seas on the planet). While we were near St. George Island in 2011, the *Europa* was drifting in strong winds and about to be shipwrecked. Nonetheless, soon afterwards the idea arose of creating a super luxury hotel in the islands, in view of the climatic trend—seen as a bonanza—and increased demand in the tourism sector.

Few studies have been done on the real impact of tourists in the area. No one really argues that they must be undertaken, especially when there are areas where hundreds of people at a time come down to walk along beaches, see penguins, bathe in hot springs or visit old whaling bases. Deterioration has been detected in historical structures, places become trampled, there are inappropriate trails, stressed penguins or seals (this is debatable, and there is much controversy about it) and uncontrolled waste, but the people who work there on a daily basis wonder who should pay for these monitoring studies. The Americans, for having the largest number of tourists? The Argentines and

Chileans, for promoting it the most? Both? Do we include a proportion for the English, Australians, Chinese and Germans?

We find ourselves once again with that undefined line of responsibility that is perfect for disengagement: the position of 'Well, it can be done' or 'Let someone else do it.' The task is made difficult by poor cooperation by the tour operators, cryptic about providing information, wary because they are expecting someone to bang on the table and say 'It's over,' the words so dreaded by those who see only problems in any kind of regulation.

Tourism in Antarctica is bound to grow. The biggest problem, however, is not the occasional disturbances caused by dumping, trampling or disturbing a few populations of birds or mammals. All of that can (and should) be regulated with an iron hand. It is not so complicated, but it requires international consensus (and should be extrapolated to the mismanagement of many bases in precarious conditions). What comes with tourism (and with much of the material and personnel of the bases) is a silent type of danger that is advancing imperceptibly all over the planet, creating huge problems everywhere: the biological invasion.

The Real Problem: Invasive Species

One of the planet's greatest extinctions occurred 375 million years ago, in the Devonian period. Thousands of species disappeared, some of which had been responsible for the very structure of their environment, creating shelter, food and complexity. One such species was coral, which did not reappear for more than a hundred million years. Some specialists blame the great extinction not on the presence of a comet or an extraterrestrial bolide, as with the dinosaurs 65 million years ago, but on certain changes that caused a flow of fauna from one place to another and promoted invasions by a few species, which subsequently took over the seas of the planet, extinguishing everything before them.

Being made extinct due to the intrusion of new species is a recurring theme in the history of our planet. Nowadays, biological invasions are one of the most serious problems ignored by society. In the course of the last century we have accelerated the exchange of species of all kinds to such an extent that it is fairly normal to be bitten by a tiger mosquito while on a terrace in Barcelona. In Antarctica, the problem associated with the flow of humans comprises the 200 species that have been unintentionally introduced into various parts of the Peninsula and adjacent archipelagos.

In 2011 I could see the result of that invasion and an increase in air temperature, as it was evident: beautiful green meadows around the Argentinean base of Jubany. It is simply impossible to control the arrival of spores, seeds, plant fragments associated with the tourists, and not even with the workers who come to the scientific bases. Life sneaks in everywhere, seeks new ways to conquer space, taking advantage of vectors such as human beings themselves to reach places it has never reached before.

The only real barrier to these new species is the cold, the harsh climatic conditions of the white continent. But in its northernmost part this pristine place is letting down its guard. As we have already said many times, there is a palpable warming in the Peninsula and especially in the islands that surround it and those forming the Scotia Arc. There, the presence of proliferating species is more acute. Weeds such as *Poa trivialis* have already settled and thrive in several areas, pushing more frugal, endemic species into strongholds in more isolated areas. The invading plants have been transported from South Africa, Great Britain and Australia (proven through population genetics) and, even after lying dormant for 284 days at −1.5 °C, have had no problem germinating, growing, being pollinated and dispersed by wind. The propagules have traveled in a short time vast distances never before experienced by any plant, by means of ships and planes: more than 16 thousand kilometers. Only a few birds are able to carry any seed on their legs across that distance, but in any case it's pointless for them to bother with the southern continent: their clumsy-winged emulators (airplanes) can reach even farther than ships, on a long journey calling in at places where people's boots, jackets and even hair can pick up and give a ride to a seed, a spore or an insect larva. On one Antarctic base, more than 40 thousand seeds were counted on a single bulldozer that had been brought in for construction work. Twenty-four different species of lichens have been detected on the Polish base (and not because it's Polish: it's just where this study was undertaken, plain and simple) on timbers brought in for construction. Genetic and phenotypic plasticity (those expressed traits that make us different from each other) is the key to the survival of many species, patiently waiting their chance to make it in the new *Terra incognita*. The rapid regression of ice and milder temperatures are leaving the way clear for new conquests.

Other invaders are more aggressive. Rats and mice have taken up residence on many of the islands of the Scotia Arc. These rodents have always been troublesome invaders. Although it is true that they partly control the new plant pests, as studies on their diet show, they also attack the chicks of species such as albatrosses or fulmars in their nests (making up to 18% of rats' diet). Mice have already transformed ecosystems on sub-Antarctic islands such as South Georgia and, if there is not far stricter control, sooner or later they will reach areas closer to the mainland itself. A direct impact on the various strata of the food chain from their predation, changes in nutrient cycles and disease vectoring are some of the problems that already need to be addressed. Once again, it seems that cold is the only regulator of mice populations. It goes from about 300 mice per hectare in summer to less than 15 in the middle of winter. But the trend is slowly reversing; mice, if they have enough food, can survive these low temperatures, increasing their numbers. The process is accelerated, and the species can travel by boat and is already reaching the white continent. Invertebrates are out of control, so have not been eradicated—even terrestrial plants have been easier to limit.

Although these terrestrial pests might be able to controlled somewhat (especially the small mammals), their parasites, bacteria and viruses cannot. What may seem harmless enough to us can be lethal for entire populations of penguins or seals. And even less controllable are those that come by sea, especially at great depths. "The evidence for species introduction by sea is more complicated than by land," says Arlie McCarthy of the University of Cambridge, Great Britain, "but species have already been detected 'waiting' for a chance to break through thermal barriers." In 2007, a well-established population of king crabs ('King Krabs'?) was detected at the base of the continental shelf of the Peninsula. They, too, are biding their time until the bottom temperatures rise high enough to exceed the 0 °C they so desperately need to survive. As on land, invasions by sea are silent, imperceptible and unstoppable. In the Antarctic, the real problem associated with the influx of humans, of whatever kind, are those species that enter unintentionally, invited by nature lovers who are not directly to blame for such a mess. While the relevant 'authorities' do have a say, in this case these bodies seem like the headless horseman of Sleepy Hollow—no head to turn to for clear and forceful action.

20

The Antarctic Treaty

Something Unique in the History of Mankind

In 2009, the so-called Antarctic Treaty turned fifty years old. I regard it as one of the few truly decent documents to have been created during the past century by multiple countries with highly diverse interests and policies. Along with the Constitution of the United Nations or the well-intentioned creation of the European Union, the Treaty of the white continent proves that, when human beings really want to agree on something, they can do it.

Signed in Washington in 1959 and put into effect in 1961, the Treaty deals simply and elegantly with the future of an expanse of sea and land ice that, due to its vast resources, could have been disputed as a potentially exploitable place. "There are only 14 articles capable of providing essential management guidelines for an entire continent, more than 10% of the planet's surface," says Paul Berkman of the Scott Polar Research Institute. Originally only 12 countries signed it, yet it must be understood that in the context of the open wound of the post-Cold War period and the marked polarization of interests across the planet, the fact that these countries were or had been enemies, such as Japan, the Soviet Union and the United States, shows it to be a unique document.

Perhaps most importantly, the Treaty makes it clear that the area below 60°S is to be used only for peaceful and scientific purposes. "Many people criticized the treaty from the 1970s and 1980s onwards for being too focused on something aseptic, like science," says Klaus Dodds of the University of London. In fact, cooperation between scientific groups from different parts of the world, with joint theoretical and practical bases, is the key to

© The Author(s), under exclusive license to Springer Nature Switzerland AG 2022

S. Rossi, *A Journey in Antarctica*, Springer Praxis Books,
https://doi.org/10.1007/978-3-030-89492-4_20

understanding this document. It has been adding new signatories, and now has a total of 47 (representing more than 80% of the planet's population).

The Treaty came about in a particularly prolific year, in terms of contacts and proposals: the International Polar Year of 1957/58. This was a time when a new perspective was proposed for the southernmost continent of the planet. As well as a place of adventure, it became seen as the last frontier, an ideal place for science in practically all its disciplines. Countries as diverse as Japan and the United States were to sit down to advance a unique consensus to unite the two (and others) in a peaceful process in which no one would be either harmed or benefited.

It is easy to think that, in reality, the Treaty came about because of the impossibility of approaching Antarctica from a technological point of view. As I have said in previous chapters, the Antarctic continent offers everything but the facilities for its exploitation, yet the truth is that in certain areas a certain degree of exploitation might have been possible, as for the whaling or fur industry in the Peninsula area. That is why it is highly commendable that, at a time when we were facing none of our now widespread globalization (given the conditions of 'factions', which had become entrenched in the United States and the Soviet Union) or serious environmental problems (at least, few people perceived them as critical, as now), the foundations of mutual respect were laid that would allow us to keep the continent out of any possible exploitation. But things are changing, balances are shifting and so are human needs (Fig. 20.1).

Unfortunately, most people are unaware of the Antarctic Treaty and its relevance to their lives. Reading these lines, you need to be aware that Antarctica is also yours. As I said earlier, it is too far away to seem to matter to us. The Treaty is important not because it protects an icy and remote place but because it involves unprecedented human cooperation. It is something akin to the space race, where several countries realized as early as the 1970s that it was far better to go together, for greater achievements, than to go separately. "Today we hardly understand what it meant that such an important treaty was signed," comments John Sears of the British Antarctic Survey (BAS). "Such cooperation in the Antarctic has allowed the discovery of the hole in the ozone layer by BAS (and its consequences on a planetary level) and the thorough recording of glacier retreat in different places." Many agree with this view: "The Antarctic Treaty has allowed information to flow between scientists from all over the planet, has diminished suspicions and promoted expeditions and even the sharing of bases in different places," explains Danila Liggett from the University of Canterbury in New Zealand.

Fig. 20.1 TV grab, especially designed to pick up intact sediment/organism samples and explore the bottoms

It was not until 1991 that the environment became a priority within the Treaty itself. The foundations were laid by reforming guidelines and protocols, with its standards the toughest in practically the whole world. It is a place free from radiation, nuclear explosions and toxic spills. As we have seen, contamination is present, yet the continent is one of the places that continues to retain the privilege of being forgotten in places where men and women crowd together (though it would undoubtedly improve their lives). On the basis of environmental cooperation, living resources seem more protected even than mineral resources. However, due to human complexity and the loopholes that we are able to find whenever something does not meet our needs, we can cheat when it suits us, even in a place as apparently protected as the southern continent.

In 2003, in the middle of a scientific cruise aboard *Polarstern*, the Germans lacked their government ministry's environmental permission to use an important tool, the hydrosweep. This is a device that, by means of sonar, collects the necessary data to 'draw' underwater maps. So far, so good. The problem is that it has been proven to affect cetaceans such as whales, dolphins and pilot whales. It is powerful even at a distance of several kilometers. The icebreaker requested permission to fly over the area first to detect any whales in order to avoid them, so as not to force them to alter course or otherwise disturb them; however, for the German Ministry of the Environment, this was not enough: the hydrosweep simply could not be used. The scientists, desperate because they knew the absurdity of the arrangement (I repeat, there were no whales in the surrounding waters at this time), did not know what to do. The group of Spanish scientists asked for permission from the Spanish Ministry of Environment. Spain also had signed the Antarctic Treaty. And this ministry granted the necessary permission. With the authorizing document in their hands, the *Polarstern* scientists could now make these valuable underwater maps.

This example shows that when we want to, we cheat—even if it is for the best of purposes. Anyone on the sidelines with bad intentions could start to do things that are not stipulated in the Treaty, including resource exploitation.

In 1999, Australia and New Zealand denounced how some signatory countries of the Treaty were exploiting fishing grounds, quickly bringing long-lived species to the point of collapse (as we have seen in previous chapters). They demanded greater coordination to prevent plundering and investment of some US$380 million for surveillance in the area. You always wonder if this sort of effort is in a good cause or to preserve a resource that has not yet been legally exploited and that some countries may consider theirs already, before any talk of real sharing. "The decisions we make in the next decade are going to be fundamental for the future of the white continent, and therefore for all of us," says Stephen Rintoul of the Commonwealth Scientific and Industrial

Research Organisation. Many think that in this decade we will have to make decisions within the framework of the Treaty—but who will be in the driver's seat?

Who Owns Antarctica?

The only place on the planet where there have been no wars is Antarctica. There have been no battles, no conquests and no reconquests, and no landings to occupy or liberate anything. Will we be able to maintain this attitude for long?

By 1972 there had already been a change in attitude toward Antarctica. It was nothing major, really, but it left a certain state of uneasiness in those (few) who were monitoring the health of the Treaty and its components. As it reads, the Treaty makes clear that the territory belongs to everyone and to no one, meaning that if someone decides to start mining, who is to stop them, de facto? One country in particular had sent a up trial balloon regarding the exploitation of 'black gold' in the area of the Peninsula at that time: Brazil. There was no specific territorial claim, just a vague desire to prospect and quantify the costs and benefits of oil extraction. This intention came to nothing, but it reinforced tensions over the so-called partition of Antarctica, a virtual map on which countries mark the sectors in which they claim a part of the territory, just in case one day the rulers sit down to deal with the issue of property of that place.

Whether they wanted or not, scientists have always been the first to enter into a new venture, the 'tip of the sword' of potential territorial claims on the white continent. I have no doubt that altruism will end at the moment that someone puts out more than a vague intention to own a part of Antarctica. The existence of bases is there to indicate, 'We are here, we have invested a lot of money and now I get a share of the exploitation.' We are gradually accepting the intrusion into the white continent. It is a gradual change, sometimes imperceptible, but one that we are viewing in slow motion, pushed by our material needs. The case of tourism, which we discussed earlier, is paradigmatic. We don't know how long this idea of an idealized Antarctica will continue; tourism is likely to peak in 2030 and there seems to be no consensus on how to deal with it.

We still consider Antarctica as a politically neutral piece of the planet, but territorial claims have always been made, even at the time when people like Amundsen or Scott aimed to achieve the glory of discovery. Chile, Argentina and Britain have overlapping claims in the area near the South American continent, even extending their potential possessions onto the continental shelf (Britain claimed another 100 thousand square kilometers in the mid-2000s) in case oil is found and can be exploited in future. Other countries such as Russia and the United States have not made any claim. They don't need to. The moment we decide to open the floodgates to exploitation, they will be the ones presiding at the table with others like China or India. There's no doubt about it. "The Antarctic Treaty is going to suffer an 'Asian' tilt soon," Liggett remarks; "they have more and more weight in the talks and in the decisions." In fact, these four countries do not recognize any territorial claims, even though other countries such as Australia or New Zealand (because of proximity), Norway (because of history and legacy) and France (I still haven't quite understood why) feel that a part of the continent should be theirs in its own right. The entry of countries with a less-westernized perspective (India or China) is seen as something natural due to their increasing influence in the world.

The partitions have been created more for strategic geographic reasons than for actual resource exploration. In fact, between 90° and 150°W nobody claims anything, perhaps because it is the farthest part from any point, including South Africa (which does not claim anything and may lose its base, due to lack of funds to maintain it). In this sense, Greenpeace and Malaysia have made a reflection that I consider very accurate: the Antarctic Treaty is a preamble to the future exploitation of the white continent. All the beauty it contains could be a dead letter in a few decades if our hungry capitalist society needs more resources, which are undoubtedly becoming scarce in many parts of the planet. Surely there will be another pocket of oil or gas, a good vein of uranium or nickel, gold and diamonds or, in any case, fishing grounds that are as yet unexploited.

In 1998 a colleague of mine witnessed at the Rothera base (owned by Great Britain) armed reconnaissance helicopters, on a continent where all kinds of weapons are supposedly prohibited. They were supposed to be a legacy of the Falklands or Malvinas War, one of so many absurd wars. It worries me that many of the inhabitants of Antarctica are military. Children born at the Argentine and Chilean Antarctic bases between the 1970s and 80s (in a puerile political stratagem of 'legitimacy' as Antarctic citizens) are now adult. It is sad to see how the dream of a unique conciliation for the whole planet is vanishing because of our eternal idiocy.

Some Further Thoughts

We could say that there are four possible scenarios for the future of the white continent. In the first, we all collaborate and we all want to preserve this pristine place. It is idealistic, but in fact it is the one that has been developed so far. In an asymmetrical framework, governments help each other, joint expeditions are undertaken and common funds are created to meet expenses. Next, the scenario may be collaborative yet also explorative. What does this mean? It means that governments collaborate and, at the same time, explore the possibilities of taking advantage of the white continent. It does not necessarily have to be all with all—it could be by bloc. For example, it could be European–United States exploitation, or Canada with Australia and New Zealand, China and other Asian countries, or with Russia. The third scenario is individual and conservationist, a 'non-aggression pact' and, at the same time, it follows independent research programs. It can last, but not forever. The fourth scenario, basically, is to let everyone do what they can and want to do.

Once again, the issue to take into account is the need for resources. In today's society, as we think of it now, this demand seems to have no limits. On the other hand, climate change is going to make Antarctica's resources more readily accessible. "It would not be bad to do a mental exercise that would allow us to see what would happen in 2070 if we continue like this, exploiting the environment as we do and causing an acceleration of climate change," reflects Rintoul. "We already know that from 1970 to 2014 we have lost 50% of the strength of the deep waters that are general in the southernmost part of the planet, we know that at this rate the melting of only the Antarctic ice will cause the sea level to rise about 27 cm by 2070 at a rate of more than 5 mm per year; we also know that if we keep going like this there will come a point where it's going to cost us no less than a trillion dollars a year in climate change impacts… and yet we look at Antarctica as a prize."

It's like what's happening in the Arctic. We are already witnessing the impacts of severe climate change, yet sea lanes are currently being established to take advantage of the melting ice. I don't think it's a good thing; we shouldn't have such a materialistic and unintelligent perspective on the profound transformations we are undergoing. Moreover, further exploitation could be highly counterproductive. Climate change could increase the capacity of the Antarctic seabed to capture 'blue carbon'. "We can consider the benthos and the organisms that inhabit it as a powerful carbon sequestering machine," says Narissa

Bax of the University of Tasmania in Australia, "but if we allow trawling in this area, we will end this ecosystem service in a very short time, as we have done in many places on the planet." It is estimated that the 'blue carbon' sequestered could alone be equivalent to about $2.27 billion. But we're missing the point. "How is it possible that Antarctica, which occupies 10% of the planet's surface, is not considered in these targets?" asks Steven Chown of Monash University in Australia. "This part of the planet has a biodiversity comparable to other places on the planet." We continue to see Antarctica as something distant, wrapped in an exemplary treaty that may cease to be so just as soon as the need arises.

21

Antarctic Bases

Scientific Occupation Strategies

When I saw the new German Neumayer station in 2011 (Neumayer III), I was puzzled. I knew its exterior has an ingenious structure that adjusted its hydraulic legs to the whims of the glacier but, accustomed to the old Georg von Neumayer II base, which was beneath the ice, the sight of that macro structure left me perplexed. The Germans strutted with restraint, proudly explaining how this culmination of high technology was designed to withstand without flinching the onslaught of one of the most challenging areas of the planet. The previous base, the old one, consisted of two enormous cylinders sunk in the ice better to insulate it from the inclement cold, especially in the winter months when the average is around -20 °C with 'peaks' of -50 or -60 °C. The new base, much larger (about 4475 m², of which 1850 m² is habitable because it is heated), would not move inexorably toward the sea. This was the fate of the old one since, being sunk in the moving ice, the 200-m deep glacier pushed it toward the edge about six miles away within a few years. The new one took this movement into account and was able to correct it, readjusting to possible irregularities in the ice. The Germans once again showed technological and monetary muscle, exhibiting accommodation for than ten people all year round in a place connected only three times a year by the *Polarstern* (and by plane at other times, if they needed to leave for a particular reason) (Fig. 21.1). And they showed it to us in the middle of the financial crisis, when in some European countries, such as Spain, we could not even honor certain social security payments due to lack of funds. It had been inaugurated in February 2009, a couple of years before my visit. When

© The Author(s), under exclusive license to Springer Nature Switzerland AG 2022 **161**
S. Rossi, *A Journey in Antarctica*, Springer Praxis Books,
https://doi.org/10.1007/978-3-030-89492-4_21

Fig. 21.1 The new Neumayer station in 2011

asked how possible cutbacks would affect its operation, the Germans mentioned that on a single round trip the *Polarstern* spent 1600 thousand € on fuel alone.

The Americans show even more muscle. They have several sites in Antarctica but their crown jewel is McMurdo, a permanent base that has a whopping 1000 inhabitants in summer and about 250 in winter. Consisting of more than a hundred barracks of various sizes, it requires a complex logistics in which three airfields (two operational only in summer), a heliport and a port play a crucial role. An icebreaker is responsible for keeping free a permanent communication channel across the Ross Sea to Ross Island (where take place, supplying its 'huge' population with food, consumables, toilet paper, chemicals, and so on. Founded in 1956, the Americans are showing off their power, making it clear that with such a facility they do not need any kind of formal territorial claim, unlike the Chileans, British or Norwegians (to give just three examples). The base has ambitious plans to build a kind of highway from there to the South Pole—a highly controversial project, since its ultimate purpose is unclear. The base hosts many research programs. It is not just another base, but it not just because it is able to hold more people that, de facto, it is the most important one. There are others that are smaller, with fewer personnel, that also conduct serious and competitive programs. Of course, there are others that do nothing or meet with major failures, as everywhere.

When you go to Antarctica and see the bases, you wonder what they are doing and why they are there. Undoubtedly, the immediate answer for those who lack innocence is the mere presence in Antarctica of a number of countries that want to be there. It is not in vain that the more than 40 facilities (some permanent and others only for the summer) of the 47 countries that have signed the Antarctic Treaty (some are shared, as we shall see) are scattered across the territory, mainly in the area of the Peninsula and its islands where there is less chance of total isolation and the logistics are simpler. The second answer is that, taking advantage of politicians' preoccupation with showing their muscle, scientists collaborate with other countries through complex multinational programs.

A good example is the base that I visited in 2000, Jubany. Located on 25 de Mayo Island, it has a permanent staff of 20 people and a fluctuating number of others, especially summer visitors. The programs are mostly Argentinean and German. Both pay costs, both extract time series, do experiments, pool knowledge and provide serious logistical support. It works. Of course, it has problems. One of them, in my opinion, is some countries' obsession with putting everything under military command as soon as they leave home. It gave the impression of a much more modest base than McMurdo but, in a sense, more efficient and less chaotic. In the environs of the base there is one site, Potter's Cove, where the effects of accelerated glacier melting on benthic fauna are monitored, finding dramatic changes in the distribution and composition of certain organisms, both in the water column and on the seafloor. Data collection and cooperation by scientists, sharing protocols and methodologies, must be the ultimate goal of the bases. At this time it is crucial, because in that part of the world (and others) very rapid changes are occurring and so an agility in exchanging is crucial. Another real utility is to test the effects of isolation for future space programs, especially at the bases that are more isolated, like Vostok or Kunlun. Maintaining the bases is essential, in my view, as are carrying out scientific expeditions and setting up temporary camps to collect data. But not at any price.

It is useless to move four or five thousand people every year (about 1000 live there in summer, of whom only 250 remain in winter, as at McMurdo) if things are not done seriously and by the appropriate means—and if the environment surrounding the base is not respected. As we have seen, one of the main problems is waste, and above all it involves the various ways in which the countries treat it according to their own protocols and budgets. It is clear that some environmental protection measures must be made more standard, otherwise some do a great deal and others much less. The Germans take their own waste away in tubular bags. In other places, the containers have gone directly into the sea, and drums full of toxic or harmful products are abandoned anywhere, even within sight.

The relationship between the inhabitants of the base and the inhabitants of the continent is more difficult. It is clear that in the first bases to be established (the Argentine and Australian meteorological bases in 1904 and 1911, in totally opposite parts of the continent), the penguins or seals were eaten in the interests of self-sufficiency (many populations were decimated). In 1990 a Hercules plane in low flight (200 m above the surface) killed 7000 penguins, crushed by an avalanche they caused in their panic. This seems a foreseeable, avoidable and punishable stupidity. The same access that the bases have from the sea to the coast has been used by creatures living there long before us, yet this is not enough for us to respect them. Here we also find asymmetries. Just as some bases are strict with both their personnel and tourists regarding access penguin colonies or the resting places for elephant seals, others encourage routine visits as a tourist attraction. The bases in Antarctica are an essential element of the future, and we need to continue to gather data and monitor the atmospheric dynamics, climate, chemical compounds in the air, species biology and effects of climate change. But the bases also represent a crucial engine of human harmony. They seem to me to be one of the few remaining bastions in which human understanding is successfully hoodwinking the politicians, who are unable to understand the importance of international cooperation, thereby ensuring that collaboration works at last.

China in Antarctica

In all this silent movement on the bases, one country that stands out for the speed with which it has progressed in barely 20 years. China has gone from having only one base (The Great Wall) in the northernmost part of the Peninsula (in one of the large archipelagos that dot this part of the world) to three, the last a real technological triumph. Kunlun is in the highest, most isolated and coldest part of the continent. It is a base designed for glaciological, paleoclimatic, atmospheric and, above all, astronomical studies. Its Dome A is undoubtedly one of the telescopes most desired by astrophysicists, who see in it a rival even to Hubble. With a diameter of 2.5 m, the 'Dark Universe' is already studying the near-infrared bands to detect planets beyond the solar system and the mysteries of dark matter. Germans,

Australians and Americans have helped on some logistical aspects, but the hard work of creating a base in possibly the most hostile place on the planet has been the task of Chinese engineers and scientists. This observatory has been difficult to design and build, yet being so high and with such a clean sky (and such a thin atmospheric layer) it will have a queue of collaborators who want to see the stars and the universe in one of the most privileged places in the world. In 2015 the plan is to have another one ready in the vanguard in these disciplines.

It is a show of strength, like the *Xue Long*, the first polar oceanographic vessel that China chartered in 2011 to demonstrate the country's abiding interest in polar research. The country's mere $10.5 million spent on scientific programs cannot be compared to the $460 million of the United States, yet the figure tripled in just 10 years and the current crisis has not dampened China's obstinacy to reach the South Pole. Why? "Polar research has become urgent for China," says Qu Tanzhou, director of the Chinese administration for the Arctic and Antarctic, in statements to the media. "We need to know more about climate change and to what extent we are responsible or not for such a phenomenon." China has surpassed the US monster in terms of emissions, and extreme weather phenomena (floods, droughts, glaciers disappearing from the Himalayas) are a reality in that country.

This Kunlun facility (built at a cost more than $150 million) will be a pioneer in this and other respects, but the Chinese openly admit that the other objective is to prospect for gas pockets, oil fields and veins of various minerals in this remote part of the planet. In the Arctic, too, China is looking for new trade routes and is willing to open a dialogue to speed up its trade to the West through waters that are currently both everybody's and nobody's territory (claimed by Canadians, Americans and Russians). In the waters surrounding the Antarctic, a 5-year program is investigating the feasibility of increasing krill fishing quotas to extract protein from the sea (China is by far the largest consumer and producer of fish on the planet).

As it is, it is not surprising to see new alliances with the Chinese, such as Australia's. The last two bases of the People's Republic of China are in territory claimed by Australia. Far from being annoyed, the Chinese and Australians are holding fluid conversations about the future of the continent, looking forward to 2048 when the closure is over and its potential exploitation will be renegotiated. In fact, the Australians opened a can of worms in 1990 by hinting at the possibility of starting serious mining prospecting, and everyone jumped on them. The 1991 Madrid Treaty was born of this strategic move, creating stricter environmental rules than even in the most supposedly civilized countries of the planet. If it is not renegotiated before, the urge to find oil and certain minerals will, as I said before, reconsider the future of this huge piece of ice with land that many consider untouchable (including myself). When resources become scarce, China and other countries will be there showing their muscle and defying the extreme climate by employing the most advanced technology in the world.

22

Living on an Ocean-Going Icebreaker

Polarstern, the World's Largest Polar Oceanographic Vessel

In this last part of the book I give a (subjective) vision of Antarctica from the point of view of a scientist voyaging to this remote place. As I have been relating throughout the chapters, I am interested in going beyond what others say, using what I have seen, what I have felt and how I see the future of the white continent.

As I mentioned in the foreword, I am not an expert on Antarctica. Undertaking three expeditions, publishing some articles and reading works of opinion and science on the subject does not give me that status. The experts are people who have been involved in the subject for up to four decades, fighting for projects, traveling and experimenting, immersed in forums where decisions are made and opinions expressed. However, I can say without hesitation that, having been there three times, having talked to people of different branches of knowledge and having gained knowhow directly (through my research) or indirectly (through the work of others) allow me a view of the whole picture in Antarctica.

Perhaps the most important thing was to go there—to see Antarctica. And, as I explained, I had the privilege of doing so aboard the best oceanographic icebreaker in the world, forged by the Germans. They have seen in the white continent and in the studies carried out there a unique opportunity to consolidate part of their political–scientific power. The *Polarstern* is the right arm of the Alfred Wegener Institute, which has appeared so many times in the

© The Author(s), under exclusive license to Springer Nature Switzerland AG 2022
S. Rossi, *A Journey in Antarctica*, Springer Praxis Books,
https://doi.org/10.1007/978-3-030-89492-4_22

stories I have been telling. Hundreds of people have attended this institution, one of the most important on the planet in terms of polar studies. Some have only passed through, while others have devoted their scientific lives to making the organization's work a reality. And most of them have had to come onboard *Polarstern* to carry out Arctic or Antarctic studies, since it is the obvious craft for such expeditions. Living on the *Polarstern* is like living in a floating hotel, with amenities, functional spaces and people at the service of science. That's why, when you have the privilege of going aboard on such a ship, you don't hesitate to enjoy it to the full.

A Day Onboard the *Polarstern*

Waking up in the morning is not often an easy matter, and when there is no distinction between Monday and Sunday the task becomes more difficult. On a scientific ship like the *Polarstern*, due to the intense work and continuous sampling, holidays as such do not exist and there comes a time when you lose track of time, asking questions like "Hey, today is Saturday, right?" The look on your cabinmate's face may enlighten you that, no, that day is not Saturday, but Monday or Tuesday. You shower, dress and are ready at 7:30 am, since the dining room closes at 8:30 am no matter what.

As you sit down you see that there is a huge amount of food that you may not need yet looks enticing: fried or scrambled eggs, a plentiful assortment of cheeses, cereals and other delicacies, all carefully prepared to get you through the day. What awaits you out there is no joke: you might have to spend a long time working on deck at several degrees below zero, and that consumes a lot of energy.

The various groups disperse from the comfort of the dining room to their laboratories to get ready: today, collecting samples or setting up underwater apparatus is on the agenda. For the men and women collecting organisms from the seabed, protection from the cold and humidity is an absolute necessity. Dressed as multicolored penguins (red, orange, black and purple), we wait for a device called an AGT (an Agassiz trawl), which is a huge iron sled carrying a trawl net to gather samples from the bottom and raise them to the surface. We are a bit anxious and excited—what will come out, will there be enough material for me? The problem with random sampling is that the various specialists dream of getting their subjects up as intact as possible and in sufficient quantity to be able to observe, treat, classify and virtually pamper them. The collection of material in Antarctica is anything but routine. That is why sometimes the preparations on deck turn into a frenzy to catch and preserve the material as soon as possible.

The classification of organisms in the various specialists' laboratories (crustaceans, fish, sponges...) is also a difficult task. Separating, labeling, photographing, freezing or fixing has to be immediate—you can't leave your material adrift. And at the start that can cause minor catastrophes: jars fall over, formalin spills, markers don't work, scissors go astray... but group synergy is soon achieved, if unexpectedly, and each nucleus of scientists seems to perform a synchronized dance that keeps everyone's efforts as efficient as possible.

In the meantime, geologists and physicists are replacing the biologists on deck. It is time to set the sediment trap with myriad devices to allow us to learn more about what happens in the water and how the suspended particles affect the organisms living at great depths according to their speed, concentration and quality. It is a challenging and tedious job, and in a few hours the generous breakfast is an inexorably fading memory. So, it's back to the dining room for lunch at 11:30 am. This seems early, but some of us appreciate it in all the hustle and bustle. One way to tell exactly what day of the week it is involves the 'floating sausage': only on Saturdays is this traditional German dish served, baffling the other nationalities.

The timetable for lunch is strict, from 11:30 to 12:30 pm. After lunch, a short coffee break will prepare us for the second wave of work. In fact, the sampling does not stop for meals, and everyone must stay attentive and not get slack through 'jaw exercise'. In any case, you can eat after hours in the sailors' mess, where a lady waits patiently for stragglers to satiate their appetite. If there is no work on deck, you can move onto laboratory work: looking at the contents of the organisms you have captured with a magnifying glass to find out what they eat and in what quantity, observing their gonads to determine their reproductive state or using a device to establish the concentration of substances such as proteins, carbohydrates or lipids in their tissues.

Time on the ship is gold, pure gold, which should never be wasted. But, yes, sometimes the sight of an iceberg, a group of penguins or a leopard seal makes the work in the lab stop. Then, equipped with cameras, video, digital, analog, or whatever, we all go on deck to enjoy the event. Enraptured by this spectacle, we are again reminded of the privilege of being here, which once the photo session is over, prompts us to return to getting results. Some have had time to have a coffee in the lounge attached to the dining room, or to read a little on one of the top decks in the small but cozy library stocked with novels, scientific books and gifts from guest scientists who may have left behind their editions in Spanish, Chinese and Indian.

The afternoon at the pole is long in spring to summer, and even after 8 or 9 pm the activity, although diminished, does not stop. While some are working out the strategy for the following day, the more physically restless go down to the pool to loosen their muscles by playing water-basketball. Drowning and blocking are the order of the day, in a game that to a large extent serves to disconnect us from the daily chores. Some prefer to contemplate the landscape through the immense glass windows of the wheelhouse, a unique place where the sheer pleasure of resting your eyes on the icy landscape makes you feel good. The light, again deceptive, invites you to stay, because the sun does not want to set and remains suspended above the ice floe until well into the night. One more day of work, and one day less before we have to go home.

Scientists Have Fun

One boat, 2 months at sea, the same people… you might think that this kind of expedition in a certain sense resembles a kind of *Big Brother* reality television show, much more civilized and rather more respectful (and certainly less idle), with over a hundred people who see each other every day and can't leave. Interestingly, the first time I heard about this show was in 2000 on the *Polarstern*… the point is that, of course, you are not scrutinizing your samples the entire time. Scientists, like the rest of us, have their moments of relaxation when they socialize outside the labs, on deck or in the canteen.

The Zillertal is a bar, and it is open on Tuesdays, Thursdays and Saturday evenings, where the sailors, officers, waitresses and scientists can have a few drinks. Unlike American ships (on which alcohol is forbidden… what poverty of spirit!), here you can share a glass of wine or drink a little beer while chatting about scientific material or, better, other things in life, tell jokes or go on at random about the first thing that comes to mind (without going overboard, because then you have to get up…). It is a necessity, because everyone needs an escape valve, a moment of communication beyond the specimen covers, wet lab or routine of analyzing samples. An increasingly magical atmosphere is created in which the ship's inhabitants get to know each other better, creating strong bonds of camaraderie and friendship that are difficult to understand if you have not lived in an enclosed space for so long. Bad feelings can also be created, but in general the synergies between individuals create good vibes of coexistence, and sometimes something more.

While fraternizing with a crew that knows how to be on the other side of the planet, observing the unique landscapes, having unrepeatable experiences and sometimes meeting unusual characters, the veterans are preparing for the rookies to 'purify' themselves as they cross the line of latitude of the Antarctic polar circle. This ritual is now approaching, and those who are to be baptized form a tight-knit bunch to rebel against those who, in the shadows, are setting a range of humiliating tests (on a ship like the *Polarstern*, none of it's obligatory). After jokes, buffoonery and other confrontations, the actual day arrives, with Neptune, Triton (his messenger) and Tethys (his wife) at the ready to witness the purification. There is a cooling-off period while the 'pollywogs' (initiates) are locked in a container on the aft deck. The next step is to cover them in mud, food scraps, flour and other debris that the veterans have been collecting for weeks prior to the ritual. After hours of this sort of behavior, everyone is exhausted and happy, celebrating over dinner the incorporation of the new 'sons of Neptune' (and daughters) who, from that moment on, are full members of the Antarctic polar community (Fig. 22.1).

By the time they arrive in port in Cape Town or Punta Arenas at the end of the polar voyage, the initiates feel like true veterans and look pityingly at the newcomers who are to take over, thus the next to campaign with the tireless *Polarstern*. There's no lack of rude comments: "It sure smells bad here… it must be the newcomers; they're going to have to purify themselves…".

Fig. 22.1 Football at Neumayer station (2003): sailors and technicians vs scientists; the first team won the match

23

The Last Bastion of an Unspoiled Planet

Feeling Isolated

In the distance I observe Comandante Ferraz, Brazil's permanent base on Deception Bay. The *Polarstern* is at anchor, motionless, waiting for a surgical operation on board to finish on a somewhat gray day. The second engineer, Ole, has suffered a sudden attack of appendicitis. The doctor on board, Ula, is operating in the ship's small operating room, assisted online from Bremen by a team of doctors. She has help from a second doctor, an Argentinean who has traveled across from the Jubany base for this purpose. It would be routine operation in a European hospital, but here the end of the world it is extraordinary.

The Brazilian base is drawn on the map as a thin yellow and green line, and without binoculars you can't tell that that's actually the colors of the flag of that country. I'm thinking all the time about Ole's operation, about how lucky they are to be on this ship and in the hands of a veteran Antarctic doctor. Ula has worked for months in isolation in Neumayer, and she is prepared to face the worst under the worst conditions. She doesn't falter as she performs the operation, despite the presence of the tormented Argentinian who talks all the time, like a machine gun. She is nervous, and it is clear that, despite her training as an expedition doctor, the situation cannot be totally under her control. This 2011 expedition has been full of minor medical setbacks. It was nothing serious, but they had to evacuate Stephi, a German student who suddenly developed a strange infection in her leg. In such situations you realize how

© The Author(s), under exclusive license to Springer Nature Switzerland AG 2022
S. Rossi, *A Journey in Antarctica*, Springer Praxis Books,
https://doi.org/10.1007/978-3-030-89492-4_23

isolated you are. Antarctica is a place from where it is difficult to get to civilization, even from the most inhabited area, the northern islands of the Peninsula (Fig. 23.1).

Already in 1967 Bryce Nelson described to us the restrictions involved in living on the white continent on a base like McMurdo where in winter 250 people were living then, rising to 1000 in summer. The technical problems were routine, and the most important issues were the frictions and neuroses being felt due to the prolonged isolation. In an article in the journal, Science, a scientist derided, "there has not been a single psychiatrist, psychologist or sociologist who has been man enough to leave his wife's bed and go to Antarctica to see what happens."

Of course, we are speaking in the past tense, for since then there have been several reviews of how behavior and synergies change in places as isolated as Neumayer. Here, a scientist can spend up to 15 consecutive months without leaving the vicinity of the base. When I hear those figures, I realize that it's no wonder that every now and then someone commits suicide, even among people who are apparently sane. Rodney Markus, of the US base Amundsen Scott, died on 12 May 2000 after ingesting large quantities of methanol, a denatured form of alcohol that is lethal in large doses. According to his friends and family it is unlikely to have been suicide. The evidence points to desperate days in a place where contact with others was restricted to those on the base. Cases much less dramatic have been repeated. The white continent is beautiful to see in passing, but is an inhospitable place in which to live. You realize

Fig. 23.1 Infinite ice, at the end of the expedition

that it is no place for humans—or for any other living being that has not undergone the millions of years of evolution necessary to adapt to the cold, the lack of light and the most absolute desolation.

When we were returning from the 2000 expedition, Wolf Arntz asked me what I missed while being on *Polarstern*: "To see something green, a tree," I replied. And it was true. I needed to see a blade of grass, a bush, a big oak tree, something to remind me that we come from a totally different ecosystem from that of the frozen Antarctic. The beauty of the ice and its bluish reflections vanish as soon as you think that you might have to see them forever.

Bouvet, the Gateway to the End of the World

If there is a remote place on the planet, it is the island of Bouvetøya, or Bouvet. The first time I saw it, I realized that it would be the perfect setting for a spy movie in which a secret organization wanted to go unnoticed. When we arrived there in February 2000 from South Africa I saw the island surrounded by a light surface mist and a crystal-clear sky above, rare in those latitudes even in the middle of the southern summer. It is of a blackish rock, crumbled into boulders and crushed by glaciers such as the Posadowsky or Christensen, which have accumulated hundreds of meters of defiant ice. Even in these splendid conditions, Bouvet exudes pure hostility: no wonder its conquest reads like a catalog of failures. When you see it in the middle of nowhere, more than 1600 km from the southernmost country in Africa or more than 2500 km from the Sandwich Islands, you realize how lonely and isolated is that piece of rock and ice.

I was not at all surprised when Wolf Arntz told me that the only flight to Bouvet Island during which he actually set foot on land was strange and extremely uncomfortable. The desolation overwhelmed him, as it must have done for Jean Baptiste Bouvet, the French explorer who sighted it in 1739. In ten days of attempts, his ship's crew were unable to set foot on land. The weather conditions, and above all the fear of being shipwrecked, must have taken their toll on that daring sailor who, at a very young age, was traveling the Atlantic in search of new communication routes and islands on which he could rest his men and ships on the endless voyages of the East India Company. Later James Cook, another pioneer of exploratory voyages, was unable to find it despite the directions of his French navigator: twice he passed close by and twice failed to locate it. It is indeed a small island, no more than 780 m high and with an area of 12 square miles. In those days, with errors of navigation due to defective cartography and the difficulty in ascertaining the latitude and longitude, it must have been like finding a needle in a haystack.

It was not until 1822 that voyagers could disembark and explore this island in even a superficial way. Benjamin Morrell was the first to see it as somewhere to stock up on seal or penguin meat, but few knew how to find it, being so far to the south and on the way to nowhere. Even the sealers and whalers feared this trap, full of danger since it is the meeting point of two of the largest oceans on the planet, the Atlantic and the Indian Oceans.

I had my first sight of the island in 2000, and in 2003 had the privilege of flying over it equipped with my cameras. The helicopter pilot gave a commentary on the island to me and two others, and on this flight the clouds and their mood changed continuously, illuminating the island with a strange light that makes it beautiful. I imagined a scene for my book, *The Fjords of Bjlaën*, a science-fiction novel in which part of the key is this remote place. My thoughts wandered over those first Norwegians who in 1927 decided to conquer it and make it theirs in 1928. It doesn't seem that they had much opposition from other countries that could potentially claim it as their own, such as France or Great Britain (South Africa), which had never had the slightest real interest in keeping it for themselves. The idea was to establish a meteorological base, but the conditions of isolation and the inclement weather aborted this plan at that time. Fifty years passed before a station was created in 1977 to detect wind, temperatures and precipitation in the prelude to the ends of the world.

And I say this because, in 2011, I realized that it really is the very edge. That year we returned to visit the 500-m cliffs of basaltic rock that plummeted, black as coal, into the waters, obscured by threatening clouds. We were returning from another long voyage and expected to reach Cape Town in a few days, and a stop at this scenery is a must because of how little it is known and how much it can tell us. Inclement weather, however, prevented us from staying more than half a day. We had to retreat in haste with our tails between our legs in view of a squall that was coming our way. These are not seas to take lightly, despite being aboard one of the most powerful scientific vessels on the planet. The impossibility of working in those conditions prompted the captain and expedition manager to cancel any plans to go ashore. Clouds surrounded the ship and the wind whipped up waves that began to rise, unhindered by any obstacle in the midst of this oceanic nothingness. Once again, the island was abandoned for further months or years with no one to visit it.

We scientists are the only ones who visit it. Why? Ute Jacob of the Alfred Wegener Institute can answer us. "Bouvet is a hostile island on its surface but very rich under the surface of the sea." Inventories of the island's fauna and flora are extremely scarce, but in 2003 a fairly exhaustive study was carried out in the short time we had available. Jacob collected a great deal of material (so did I, by the way) and analyzed the food chain of the site on its rich bottoms. "The trophic relationships are very similar to those in Antarctica, both in terms of the organisms there and the material they consume," says Jacob. The benthos is teeming with organisms and is an oasis for many species, which adopt it as a stopover or even establish colonies there. In fact, any island is an oasis. By intercepting currents and creating zones where benthic fauna are abundant in the middle of abyssal flats, an island causes life to congregate to feast on the produce of its waters, often rich. In Bouvet this detail is significant, and it is a place of passage for many creatures, which do not see it as a hostile place but as another chance to survive.

Not all creatures stay. A young colony of chinstrap penguins was seen to have transferred from Bouvet to the Sandwich Islands across the Atlantic Ocean, swimming more than 3600 km, day and night. "The penguins were moving at about 2.5 m per second," says Martin Biuw of the Norwegian Polar Institute, who studied their movements using satellite imagery. "The movement was not linear, that's not the minimum distance between the two points," Biuw adds, "but it's that the penguins are guided by ocean currents, by the associated food and by the minimum effort they have to make to get to their destination." Like airplanes along many commercial routes, they did not follow the shortest way to their destination but the one with the strongest currents, economizing on fuel.

Every time I leave Bouvet either to go to or return from the white continent, I realize how lucky I am to witness something so unique. It is one of those areas where you know few eyes have rested or scrutinized. That is why, when I was given the opportunity to fly over it by helicopter, I felt the luckiest person on the planet.

Visions of the Last Bastion

I can't imagine Ui-Te-Rangiora's face when, around 650AD, having set out from New Zealand with a group of Polynesian navigators, he saw the immense walls of Antarctica. These early visitors came to see icebergs—at least, that is what is inferred from the oral legends of their islands. They possibly reached the shores of the white continent before returning and telling their people that there was nothing good for them there. Consummate navigators, the opportunity to conquer new territories moved them to take advantage of the currents during long journeys in which they went for weeks without landing. Seeing those ice cathedrals must have both impressed and frightened them. I suppose that, like the European navigators many centuries later, they understood that this was the true end of the world.

Antarctica leaves an indelible mark on anyone who sees it for the first time. There is nothing that can compare, nothing so devoid of life and at the same time so beautiful. It provokes contradictory feelings when contemplating it. Not even a place as remote and cold as Greenland produces as much uneasiness and pleasure as the immense Antarctic icebergs, because in the Arctic there is land, there are people, there is green. In Antarctica, apart from the area of the Peninsula (and only in four very specific places), the ice and the occasional rock on the coast show a landscape never conquered or fully assimilated by humans (Fig. 23.2).

For this reason, when in the cockpit of the *Polarstern* we talked about the unique opportunity we had been given to reach places that possibly no one had ever seen, we were also aware that such a place cannot be understood only through photos and documentaries. I was lucky enough to visit it three times, and watching the iridescent halos and austral auroras forming green curtains across the sky like a strange, undulating sheet gave me the opportunity to think about things from a completely different perspective than when surrounded by a non-hostile environment.

On the ship's heliport we often had a soda or a beer while the sun set or the sea, a total calm, reflected the silhouettes of icebergs. After working on the rigging or in laboratories all day, the crew, scientific or not, would capture that

Fig. 23.2 Blue iceberg, observed at the end of the 2000 expedition

magical moment when the lights play with the ice and the surface of the sea. Antarctica is a challenge to the imagination, and describing its landscape is much more complicated than portraying a jungle, a desert or a Mediterranean coast. Because we have all seen those landscapes as human beings through conquering the entire planet in stages, reaching even remote places. Never having colonized the southernmost continent of the planet has not permitted us to leave an imprint of our genetic vision of the landscape. Writers such as Howard Phillips Lovecraft, Edgar Alan Poe and Jules Verne have used this place to immerse their characters in impossible locations, where the landscape surpasses comprehension by the protagonists, who find themselves amazed and at the same time terrified before something that they do not understand. "Antarctica is too much for any human being," Lloyd Peck of the British Antarctic Survey rightly points out, in my opinion, "perhaps because it is a place out of the ordinary, out of the everyday."

Antarctica is something else. When you have studied it and realized that we only know superficial things about its constitution, functioning and role within our planet, you appreciate that it is impossible to explain everything through the scientific prism alone. The white continent is the clearest example that, separately, the various branches of science will never reach their goal of understanding. Only a holistic view, of the whole, can allow us to make even a superficial approach to its complexity. A few blocks of ice and rock hide the fabulous diversity under its waters, and one of the harshest climates on the planet supports a quantity of life that is scarcely comprehensible if we do not understand the parts of the puzzle that combine for it to survive and proliferate exuberantly in the southernmost and most hostile seas of the world. Being the most distant, the landscapes and changes in the Antarctic are the least understood.

As the philosopher Slavoj Žižek said, "Society must change and stop seeing nature as a good and involved mother; it must see it as a pale and indifferent mother." Only the vision of the whole allows us to have that vision, and the landscapes of Antarctica are unique in making us understand that we are small, very small. And that after our passage, nature will continue unperturbed on its path, spinning, after hundreds of millions of years of victories and failures that no one has ever celebrated because there is nothing to celebrate.

Epilogue

The question is, so what? Do we care about what happens in Antarctica? I think that the vast majority of readers, after this brief review of such a vast subject encompassing an entire continent, may have asked and answered this question. Let's not fool ourselves. We can be sensitive, look to improve things, take an interest, apply our intellect to create opinion... but if something that happens nearby is already too removed for us to address it coherently, Antarctica is certainly too distant for us to be able to act in a straightforward, simple way.

We live in a society where almost half of the population lives in a medium or large city. Nature is far away from us. That weekend we spent in the mountains or at a secluded cove on the coast is not enough to understand and assimilate the sensitivity we require to react and integrate ourselves into a world more in line with our possibilities. That is why it is so hard for us to change things. We do not understand how dependent we are on our environment, because everything comes to us pre-manufactured, from far away. I am the first to admit this problem, don't worry. No matter how much I study, travel, or become involved in conservation issues, my life continues to roll forward on its complex wheel, oblivious to nature and the environment that sustain me.

This book has been an attempt to perform a complete about-face. I wanted to bring to those who want to know a place of indescribable beauty, virgin territory in many aspects, the last frontier of an anthropized planet in which a single species is managing to change its climate, the carbon and its very life flows, the interactions between organisms and the number of species that

© The Author(s), under exclusive license to Springer Nature Switzerland AG 2022
S. Rossi, *A Journey in Antarctica*, Springer Praxis Books,
https://doi.org/10.1007/978-3-030-89492-4

populate it, in an apparent quest to grow without limit. Hopefully, Antarctica will be seen now as something more than an accumulation of penguins and seals, as a fortress of ice that only a privileged few will visit in their lifetime, some at the taxpayer's expense. Hopefully, this book will serve to bring us closer to a fascinating and mysterious continent that still has the taste of adventure, which for an old species like us most of the planet has lost, which no longer surprises us. And hopefully, when decisions begin to be made, some of you, if you have studied issues such as climate change, mining or fishing in Antarctica or ecotourism on the white continent, will choose to answer, "I do care about what happens down there, because it will affect me **directly**, even if I am 12,500 kilometers away".

The Springer Praxis Popular Science series contains fascinating stories from around the world and across many different disciplines. The titles in this series are written with the educated lay reader in mind, approaching nitty-gritty science in an engaging, yet digestible way. Authored by active scholars, researchers, and industry professionals, the books herein offer far-ranging and unique perspectives, exploring realms as distant as Antarctica or as abstract as consciousness itself, as modern as the Information Age or as old our planet Earth. The books are illustrative in their approach and feature essential mathematics only where necessary. They are a perfect read for those with a curious mind who wish to expand their understanding of the vast world of science.

Bibliography

P.A. Abrams, D.G. Ailey, L.K. Blight, et al., Necessary elements of precautionary management: implications for the Antarctic toothfish. Fish Fish. **17**, 1152–1174 (2016)

D.G. Ainley, P.A. Cziko, N. Nur, et al., Further evidence that Antarctic toothfish are important to Weddell seals. Antarct. Sci., 1–13 (2020)

R.B. Alley, On thickening ice? Science **295**, 451–452 (2002)

R.B. Alley, Abrupt climate changes: Oceans, ice, and us. Oceanography **17**, 194–206 (2004)

R.B. Alley, P.U. Clark, P. Huybrechts, et al., Ice-sheet and sea-level changes. Science **310**, 456–460 (2005)

R.B. Alley, M. Fahnestock, I. Joughin, Understanding glacier flow in changing times. Science **322**, 1061–1062 (2008)

P.M. Almond, K. Linse, S. Dreutter, et al., In-situ image analysis of habitat heterogeneity and benthic biodiversity in the Prince Gustav Channel, Eastern Antarctic Peninsula. Front Mar. Sci. **8**, 614496 (2021)

G. Alurralde, V.L. Fuentes, T. Maggioni, et al., Role of suspension feeders in Antarctic pelagic-benthic coupling: trophic ecology and potential carbon sinks under climate change. Mar. Environ. Res. **152**, 104790 (2019)

M.J. Amesbury, T.P. Roland, J. Royles, et al., Widespread biological response to rapid warming on the Antarctic Peninsula. Curr. Biol. **27**, 1616–1622 (2017)

W.S. Andriuzzi, B.J. Adams, J.E. Barret, et al., Observed trends of soil fauna in the Antarctic Dry Valleys: early signs of shifts predicted under climate change. Ecology **99**, 312–321 (2018)

A.M. Anesio, A.J. Hodson, A. Fritz, et al., High microbial activity on glaciers: importance to the global carbon cycle. Glob. Change Biol. **15**, 955–960 (2009)

W.E. Arntz, T. Brey, V.A. Gallardo, et al., Antarctic zoobenthos. Oceanogr. Mar. Biol. **32**, 241–304 (1994)

© The Author(s), under exclusive license to Springer Nature Switzerland AG 2022
S. Rossi, *A Journey in Antarctica*, Springer Praxis Books,
https://doi.org/10.1007/978-3-030-89492-4

W.E. Arntz, S. Thatje, D. Gerdes, et al., The Antarctic–Magellan conection: macrobenthos ecology on the shelf and upper slope, a progress report. Sci. Mar. **69**(Suppl. 2), 237–269 (2005)

R.B. Aronson, R.M. Moody, L.C. Ivany, et al., Climate change and trophic response of the Antarctic bottom fauna. PLoS One **4**, e4385 (2010)

K.R. Arrigo, C.R. McClain, Spring phytoplankton production in the Western Ross Sea. Science **266**, 261–263 (1994)

K.R. Arrigo, D.H. Robinson, D.L. Worthen, et al., Phytoplankton community structure and the drawdown of nutrients and CO_2 in the Southern Ocean. Science **283**, 365–367 (1999)

R.J. Arthern, C.R. Williams, The sensitivity of West Antarctica to the submarine melting feedback. Geophys. Res. Lett. **44**, 2352–2359 (2017)

E. Ayres, J.N. Nkem, D.H. Wall, et al., Experimentally increased snow accumulation alters soil moisture and animal community structure in a polar desert. Polar Biol. **33**, 897–907 (2010)

H. Bae, I.Y. Ahn, J. Park, et al., Shift in polar benthic community structure in a fast retreating glacial area of Marian Cove, West Antarctica. Sci. Rep. **11**, 241 (2021)

J.L. Bamber, R.E.M. Riva, B.L.A. Vermeersen, et al., Reassessment of the potential sea-level rise from a collapse of the west Antarctic ice sheet. Science **324**, 901–903 (2009)

S. Ban, N. Ohi, S.C.Y. Leong, et al., Effect of solar ultraviolet radiation on survival of krill larvae and copepods in Antarctic Ocean. Polar Biol. **30**, 1295–1302 (2007)

A. Barbosa, M.J. Palacios, Health of Antarctic birds: a review of their parasites, pathogens and diseases. Polar Biol. **32**, 1095–1115 (2009)

C. Barbraud, H. Weimerskirch, Emperor penguins and climate change. Nature **411**, 183–186 (2001)

C. Barbraud, H. Weimerskirch, Antarctic birds breed later in response to climate change. PNAS **103**, 6248–6251 (2006)

C. Barbraud, P. Rivalan, P. Inchausti, et al., Contrasted demographic responses facing future climate change in Southern Ocean seabirds. J Anim. Ecol. **80**, 89–100 (2011)

D.K.A. Barnes, Iceberg killing fields limit huge potential for benthic blue carbon in Antarctic shallows. Glob. Change Biol. **23**, 2649–2659 (2017)

D.K.A. Barnes, L.S. Peck, S.A. Morley, Ecological relevance of laboratory determined temperature limits: colonization potential, biogeography and resilience of Antarctic invertebrates to environmental change. Glob. Change Biol. **16**, 3164–3169 (2010)

D.K.A. Barnes, A. Fleming, C.J. Sands, et al., Icebergs, sea ice, blue carbon and Antarctic climate feedbacks. Philos. Trans. R. Soc. A **376**, 20170176 (2018)

D.K.A. Barnes, C.J. Sands, A. Cook, et al., Blue carbon gains from glacial retreat along Antarctic fjords: what should we expect? Glob. Change Biol. **26**, 2750–2755 (2020)

E. Barrera-Oro, The role of fish in the Antarctic marine food web: differences between inshore and offshore waters in the southern Scotia Arc and west Antarctic Peninsula. Antarct. Sci. **14**(4), 293–309 (2002)

J.E. Barret, R.A. Virginia, D.H. Wall, et al., Persistent effects of a discrete warming event on a polar desert ecosystem. Glob. Change Biol. **14**, 2249–2261 (2008)

N. Bax, C.J. Sands, B. Gogarty, et al., Increasing blue carbon around Antarctica is an ecosystem service of considerable societal and economic value worth protecting. Glob. Change Biol. **27**, 5–12 (2021)

N.A. Bender, K. Crosbie, H.J. Lynch, Patterns of tourism in the Antarctic Peninsula region: a 20-year analysis. Antarct. Sci. **28**, 194–203 (2016)

J.R. Bennett, J.D. Shaw, A. Terauds, et al., Polar lessons learned: long-term management based on shared threats in Arctic and Antarctic environments. Front. Ecol. Environ. **13**, 316–324 (2015)

M.J. Benton, Polar dinosaurs and ancient climate. Trends Ecol. Evol. **6**, 28–30 (1991)

J. Biuw, O.A. Nost, A. Stien, et al., Effects of hydrographic variability on the spatial, seasonal and diel diving patterns of southern elephant seals in the Eastern Weddell Sea. PLoS One **5**, e13816 (2010a)

J. Biuw, C. Lydersen, P.J. De Bruyn, et al., Long-range migration of a chinstrap penguin from Bouvetøya to Montagu Island, South Sandwich Islands. Antarct. Sci. **22**(2), 157–162 (2010b)

T. Blunier, J. Chappellaz, J. Schwander, et al., Asynchrony of Antarctic and Greenland climate change during the last glacial period. Nature **394**, 739–743 (1998)

A. Brandt, A.J. Gooday, S.N. Brandao, et al., First insights into the biodiversity and biogeography of the Southern Ocean deep sea. Nature **447**, 307–311 (2007)

W.S. Broecker, Does the trigger for abrupt climate change reside in the Ocean or in the atmosphere? Science **300**, 1519–1522 (2003)

C.M. Brooks, Watch over Antarctic waters. Nature **558**, 177–180 (2018)

D. Buskowiak, D. Janussen, An exceptional occurrence of deep-sea sponges in the region of former Larsen Ice Shelves, Antarctic Peninsula, with the description of two new species. Mar. Biodivers. **51**, 8 (2021)

H.A. Campbell, K.P.P. Fraser, C.M. Bishop, et al., Hibernation in an Antarctic fish: on ice for winter. PLoS One **3**, e1743 (2008)

S. Cauvy-Fraunié, O. Dangles, A global synthesis of biodiversity responses to glacier retreat. Nat. Ecol. Evol. **3**, 1675–1685 (2019)

E.L. Cavan, A. Belcher, A. Atkinson, et al., The importance of Antarctic krill in biogeochemical cycles. Nat. Commun. **10**, 4742 (2019)

A. Cazenave, How fast are the ice sheets melting? Science **314**, 1250–1252 (2006)

M.P. Chipperfield, S. Bekki, S. Dhomse, et al., Detecting recovery of the stratospheric ozone layer. Nature **549**, 211–218 (2017)

S.L. Chown, C.M. Brooks, A. Terauds, et al., Antarctica and the strategic plan for biodiversity. PLoS Biol **15**(3), e2001656 (2017)

W.L. Chu, N.L. Dang, Y.Y. Kok, et al., Heavy metal pollution in Antarctica and its potential impacts on algae. Polar Sci. **20**, 75–83 (2019)

K.J. Chwedorzewska, Poa annua L. in Antarctic: searching for the source of introduction. Polar Biol. **31**, 263–268 (2008)

K.J. Chwedorzewska, M. Korczak-Abshire, A. Znój, Is Antarctica under threat of alien species invasion? Glob. Change Biol. **6**, 1942–1943 (2019)

P.U. Clark, A.S. Dyke, J.D. Shakun, et al., The last glacial maximum. Science **325**, 710–714 (2009)

A. Clarke, Costs and consequences of evolutionary temperature adaptation. Trends Ecol. Evol. **18**(11), 573–581 (2003)

A. Clarke, I.A. Johnston, Evolution and adaptive radiation of Antarctic fishes. Trends Ecol. Evol. **11**, 212–218 (1996)

A. Clarke, D.K.A. Barnes, D.A. Hodson, How isolated is Antarctica? Trends Ecol. Evol. **20**(1), 1–3 (2005)

K.E. Conlan, S.L. Kim, H.S. Lenihan, et al., Benthic changes during 10 years of organic enrichment by McMurdo Station, Antarctica. Mar. Pollut. Bull. **49**, 43–60 (2004)

K.E. Conlan, S.L. Kim, A.R. Thurber, et al., Benthic changes at McMurdo Station, Antarctica following local sewage treatment and regional iceberg-mediated productivity decline. Mar. Pollut. Bull. **60**, 419–432 (2010)

A.J. Constable, J. Melbourne-Thomas, S.P. Corney, et al., Climate change and Southern Ocean ecosystems I: how changes in physical habitats directly affect marine biota. Glob. Change Biol. **20**, 3004–3025 (2014)

P. Convey, L.S. Peck, Antarctic environmental change and biological responses. Sci. Adv. **11**, eaaz0888 (2019)

P. Convey, G. Bindschadler Di Prisco, et al., Antarctic climate change and the environment. Antarct. Sci. **21**, 581–563 (2009)

H.E. Copeland, K.E. Doherty, D.E. Naugle, et al., Mapping oil and gas development potential in the US Intermountain West and estimating impacts to species. PLoS One **4**, e7400 (2009)

M.P. Correa, A.L.C. Yamamoto, G.R. Moraes, et al., Changes in the total ozone content over the period 2006 to 2100 and the effects on the erythemal and vitamin D effective UV doses for South America and Antarctica. Photochem. Photobiol. Sci. **18**, 2931 (2019)

R. Craawford, U. Ellenberg, E. Frere, et al., Tangled and drowned: a global review of penguin bycatch in fisheries. Endang. Species Res. **34**, 373–396 (2017)

J.P. Croxall, S. Nicol, Management of Southern Ocean fisheries: global forces and future sustainability. Antarct. Sci. **16**(4), 569–584 (2004)

R. Cullen, Antarctic minerals and conservation. Ecol. Econ. **10**, 143–155 (1994)

J.P. Darling, D.D. Garland, L.F. Stanish, et al., Thermal autecology describes the occurrence patterns of four benthic diatoms in McMurdo Dry Valley streams. Polar Biol. **40**, 2381–2396 (2017)

H. De Angelis, P. Skvarca, Glacier surge after ice shelf collapse. Science **299**, 1560–1562 (2003)

S.L. Deppeler, A.T. Davidson, Southern Ocean phytoplankton in a changing climate. Front. Mar. Sci. **4**, 40 (2017)

H.M. Dierssen, R.C. Smith, et al., Glacial meltwater dynamics in coastal waters west of the Antarctic península. PNAS **99**, 1790–1795 (2002)

G.R. DiTullio, J.M. Grebmeier, K.R. Arrigo, et al., Rapid and early export of Phaeocystis Antarctica blooms in the Ross Sea, Antarctica. Nature **404**, 595–598 (2000)

K. Dodds, A.D. Hemmings, P. Roberts, *Handbook on the Politics of Antarctica* (Edward Elgar, Northampton, MA, 2017), 630 pp

E. Domack, D. Duran, A. Leventer, et al., Stability of the Larsen B ice shelf on the Antarctic Peninsula during the Holocene epoch. Nature **436**, 681–685 (2005)

M.B. Dyurgerov, M.F. Meier, Twentieth century climate change: evidence from small glaciers. PNAS **97**, 1406–1411 (2000)

J.T. Eastman, The nature of the diversity of Antarctic fishes. Polar Biol. **28**, 93–107 (2005)

F. Elias-Piera, S. Rossi, J.M. Gili, et al., Trophic ecology of seven Antarctic gorgonians. Mar. Ecol. Prog. Ser. **477**, 93–106 (2013)

F. Elias-Piera, S. Rossi, M.A.V. Petti, et al., Fauna associated with morphologically distinct macroalgae from Admiralty Bay, King George Island (Antarctica). Polar Biol. **43**, 1535–1547 (2020)

S.Z. El-Sayed, History and evolution of primary productivity studies of the Southern Ocean. Polar Biol., 28423–28438 (2005)

M.R. England, T.J.W. Wagner, I. Eisemnan, Modeling the breakup of tabular icebergs. Sci. Adv. **6**, eabd1273 (2020)

J.T. Ensminger, L.N. McCold, J.W. Webb, Environmental impact assessment under the National Environmental Policy Act and the protocol on environmental protection to the Antarctic Treaty. Environ. Manage. **24**, 13–23 (1999)

J.A. Ericson, N. Hellessey, S. Kawaguchi, et al., Near-future ocean acidification does not alter the lipid content and fatty acid composition of adult Antarctic krill. Sci. Rep. **9**, 12375 (2019)

V.J. Fabry, J.B. McClintock, J.T. Mathis, et al., Ocean acidification at high latitudes. Oceanography **22**, 161–171 (2009)

L.V. Ferrada, Five factors that will decide the future of Antarctica. Polar J. **8**, 84–109 (2018)

R. Ferrari, M.F. Jansen, J.F. Adkins, et al., Antarctic sea ice control on ocean circulation in present and glacial climates. PNAS **111**, 8753–8758 (2014)

G.E. Fogg, *The biology of polar hábitats* (Oxford University Press, 1998), 263 pp

J. Forcada, P.N. Trathan, Penguin responses to climate change in the Southern Ocean. Glob. Change Biol. **15**, 1618–1630 (2009)

D. Fox, Could East Antarctica be headed for big melt? Science **328**, 1630–1631 (2010)

V. Fuentes, G. Alurralde, B. Meyer, et al., Glacial melting: an overlooked threat to Antarctic krill. Sci. Rep. **6**, 27234 (2016)

O. Gagliardini, The health of Antarctic ice shelves. Nat. Clim. Change **8**, 14–21 (2018)

D. Gerdes, E. Isla, R. Knust, et al., Response of Antarctic benthic communities to disturbance: first results from the artificial Benthic Disturbance Experiment on the eastern Weddell Sea Shelf, Antarctica. Polar Biol. **31**, 1469–1480 (2008)

J.M. Gili, S. Rossi, F. Pagés, et al., A new link between the pelagic and benthic systems in the Antarctic shelfs. Mar. Ecol. Prog. Ser. **322**, 43–49 (2006a)

J.M. Gili, W.E. Arntz, A. Palanques, et al., A unique assemblage of epibenthic sessile suspension feeders with archaic features in the high-Antarctic. Deep-sea Res. Part II **53**, 1029–1052 (2006b)

J.M. Gili, C. Orejas, E. Isla, et al., Seasonality on the high Antarctic benthic shelf communities? in *Antarctic climate change and the environment. ACCE report*, ed. by J. Turner, P. Convey, G. di Prisco, P. Mayewski, D. Hodgson, E. Fahrbach, B. Bindschadler, (Cambridge University Press, Cambridge, 2009), pp. 276–278

F. Gillet-Chaulet, G. Durand, Ice-sheet advance in Antarctica. Nature **467**, 794–795 (2010)

C.H. Greene, D.J. Baker, D.H. Miller, A very inconvenient truth. Oceanography **23**, 214–218 (2010)

J. Grieger, G.C. Leckebusch, U. Ulbrich, Net precipitation of Antarctica: thermodynamical and dynamical parts of the climate change signal. J. Clim. **29**, 907–924 (2016)

H.J. Griffiths, Antarctic marine biodiversity – what do we know about the distribution of life in the Southern Ocean? PLoS One **5**, e11683 (2010)

L. Gross, As the Antarctic ice pack recedes, a fragile ecosystem hangs in the balance. PLoS Biol **3**, e127 (2005)

C. Gura, S.O. Rogers, Metatranscriptomic and metagenomic analysis of biological diversity in Subglacial Lake Vostok (Antarctica). Biology **9**, 55 (2020)

J. Gutt, Coexistence of macro-zoobenthic species on the Antarctic shelf: an attempt to link ecological theory and results. Deep-Sea Res. II **53**, 1009–1028 (2006)

J. Gutt, I. Barratt, E. Domack, et al., Biodiversity change after climate-induced ice-shelf collapse in the Antarctic. Deep-Sea Res. Part II **58**, 74–83 (2010)

J. Gutt, V. Cummings, P.K. Dayton, et al., Antarctic marine animal forests: threedimensional communities in Southern Ocean ecosystems, in *Marine animal forests: the ecology of benthic biodiversity hot - spots*, ed. by S. Rossi, L. Bramanti, A. Gori, C. Orejas, (Springer, Cham, 2017), pp. 315–344

J. Gutt, J. Arndt, C. Kraan, et al., Benthic communities and their drivers: a spatial analysis off the Antarctic Peninsula. Limnol. Oceanogr. **64**, 2341–2357 (2019)

H. Han, S. Lee, J.I. Kim, et al., Changes in a giant iceberg created from the collapse of the Larsen C Ice Shelf, Antarctic Peninsula, derived from Sentinel-1 and CryoSat-2 Data. Remote Sens. **11**, 404 (2019)

A.M. Hancock, C.K. King, J.S. Stark, et al., Effects of ocean acidification on Antarctic marine organisms: a meta-analysis. Ecol. Evol. **10**, 4495–4514 (2020)

C.M. Harris, Aircraft operations near concentrations of birds in Antarctica: the development of practical guidelines. Biol. Conserv. **125**, 309–322 (2005)

J.T. Hinke, A.M. Cossio, M.E. Goebel, et al., Identifying risk: concurrent overlap of the Antarctic Krill fishery with krill-dependent predators in the Scotia Sea. PLoS One **12**(1), e0170132 (2017)

L. Hoffmann, S.L. Eggers, E. Allhusen, et al., Interactions between the ice algae Fragillariopsis cylindrus and microplastics in sea ice. Environ. Int. **139**, 105697 (2020)

A.E. Hogg, G.H. Gudmundsson, Impacts of the Larsen-C ice shelf calving event. Nat. Clim. Change **7**, 540–542 (2017)

J.T. Hollibaugh, C. Lovejoy, A.E. Murray, Microbiology in Polar Oceans. Oceanography **20**, 140–145 (2007)

L.A. Hückstädt, M.D. McCarthy, P.L. Koch, et al., What difference does a century make? Shifts in the ecosystem structure of the Ross Sea, Antarctica, as evidenced from a sentinel species, the Weddell seal. Proc. R. Soc. B **284**, 20170927 (2017)

L.A. Hückstädt, A. Piñones, D.M. Palacios, et al., Projected shifts in the foraging habitat of crabeater seals along the Antarctic Peninsula. Nat. Clim. Change (2020). https://doi.org/10.1038/s41586-020-2126-y

K.A. Hughes, J.E. Lee, C. Ware, et al., Impact of anthropogenic transportation to Antarctica on alien seed viability. Polar Biol. **33**, 1125–1130 (2010)

Hughes KA, Fretwell P, Rae J et al (2011) Untouched Antarctica: mapping a finite and diminishing environmental resource. Antarct. Sci. 1 of 12

K.A. Hughes, L.R. Pertierra, M.A. Molina-Montenegro, et al., Biological invasions in terrestrial Antarctica: what is the current status and can we respond? Biodivers. Conserv. **24**, 1031–1055 (2015)

D.A. Hutchins, P.W. Boyd, Marine phytoplankton and the changing ocean iron cycle. Nat. Clim. Change **6**, 1072–1079 (2016)

P. Huybers, Antarctica's orbital beat. Science **325**, 1085–1086 (2009)

B. Imbert, *North pole, South pole: journeys to the ends of the Earth* (New Horizons, New York, 1992), 191 pp

J. Ingels, R.B. Aronson, C.R. Smith, et al., Antarctic ecosystem responses following ice-shelf collapse and iceberg calving: science review and future research. WIREs Clim. Change **12**, e682 (2020)

O. Ingolfsson, C. Hjort, P.A. Berkman, et al., Antarctic glacial history since the last glacial maximum: an overview of the record on land. Antarct. Sci. **10**(3), 326–344 (1998)

E.R. Irvins, Ice sheet stability and sea level. Science **324**, 888–889 (2009)

E. Isla, DeMaster, Labile organic carbon dynamics in continental shelf sediments after the recent collapse of the Larsen ice shelves off the eastern Antarctic Peninsula: a radiochemical approach. Geochimica et Cosmochimica Acta **242**, 34–50 (2018)

E. Isla, S. Rossi, A. Palanques, et al., Biochemical composition of the sediment from the Eastern Weddell Sea High nutritive value in a high benthic-biomass environment. J. Mar. Syst. **60**, 255–267 (2006)

E. Isla, D. Gerdes, S. Rossi, et al., Biochemical characteristics of surface sediments on the eastern Weddell Sea continental shelf, Antarctica: is there any evidence of seasonal patterns? Polar Biol. **34**, 1125–1133 (2011)

E. Isla, E. Pérez-Albadalejo, C. Porte, Toxic anthropogenic signature in Antarctic continental shelf and deep sea sediments. Sci. Rep. **8**, 9154 (2018)

U. Jacob, T. Brey, I. Fetzer, et al., Towards the trophic structure of the Bouvet Island marine ecosystem. Polar Biol. **29**, 106–113 (2006)

L.J. Janiot, J.L. Sericano, O. Marcucci, et al., Evidence of oil leakage from the Bahia Paraiso wreck in Arthur Harbour, Antarctica. Mar. Pollut. Bull. **46**, 1615–1629 (2003)

J. Jansen, N.A. Hill, P.K. Dunstan, et al., Abundance and richness of key Antarctic seafloor fauna correlates with modelled food availability. Nat. Ecol. Evol. (2017). https://doi.org/10.1038/s41559-017-0392-3

S. Jenouvrier, J. Garnier, F. Patout, et al., Influence of dispersal processes on the global dynamics of Emperor penguin, a species threatened by climate change. Biol. Conserv. **212**, 63–73 (2017)

M.G.W. Jones, P.G. Ryan, Evidence of mouse attacks on albatross chicks on sub-Antarctic Marion Island. Antarct. Sci. **22**(1), 39–42 (2010)

J. Jouzel, J.R. Petit, R. Souchez, et al., More than 200 meters of lake ice above sub-glacial Lake Vostok, Antarctica. Science **286**, 2138–2141 (1999)

H. Joy-Warren, G. van Dijken, A.C. Alderkamp, et al., Light is the primary driver of early season phytoplankton production along the Western Antarctic Peninsula. J. Geophys. Res.: Oceans *124*, 7375–7399 (2019)

S.E. Karelitz, S. Uthicke, S.A. Foo, et al., Ocean acidification has little effect on developmental thermal windows of echinoderms from Antarctica to the tropics. Glob. Change Biol. **23**, 657–672 (2017)

D. Karentz, Ecological considerations of Antarctic ozone depletion. Antarct. Sci. **3**, 3–11 (1991)

D.M. Karl, D.F. Bird, K. Björkman, et al., Microorganisms in the accreted ice of Lake Vostok, Antarctica. Science **286**, 2144–2147 (1999)

F. Kasamatsu, K. Matsuoka, T. Hakamada, Interspecifc relationships in density among the whale community in the Antarctic. Polar Biol. **23**, 466–473 (2000)

D.E. Kaufman, M.A.M. Friedrichs, W.O. Smith Jr., et al., Climate change impacts on southern Ross Sea phytoplankton composition, productivity, and export. J. Geophys. Res.: Oceans **122**, 2339–2359 (2017)

S. Kawaguchi, S. Nicol, A.J. Press, Direct effects of climate change on the Antarctic krill fishery. Fish. Manage. Ecol. **16**, 424–427 (2009)

A. Kelly, D. Lannuzel, T. Rodemann, et al., Microplastic contamination in east Antarctic sea ice. Mar. Pollut. Bull. **154**, 111–130 (2020)

H.J.R. Keys, S.S. Jacobs, D. Barnett, The calving and drift of iceberg B-9 in the Ross Sea, Antarctica. Antarct. Sci. **2**(3), 243–257 (1990)

H. Kim, H.W. Ducklow, D. Abele, et al., Inter-decadal variability of phytoplankton biomass along the coastalWest Antarctic Peninsula. Philos. Trans. R. Soc. A **376**, 20170174 (2018)

B.M. Kim, A. Amores, S. Kang, et al., Antarctic blackfin icefish genome reveals adaptations to extreme environments. Nat. Ecol. Evol. **3**, 469–478 (2019)

N.R. Kirkham, K.M. Gjerde, A.M.W. Wilson, DEEP-SEA mining: policy options to preserve the last frontier – lessons from Antarctica's mineral resource convention. Mar. Policy **115**, 103859 (2020)

K.H. Kock, Antarctic icefishes (Channichthyidae): a unique family of fishes. A review. Polar Biol. **28**, 862–895 (2005)

K.H. Kock, M.G. Purves, G. Duhamel, Interactions between Cetacean and fisheries in the Southern Ocean. Polar Biol. **29**, 379–388 (2006)

K. Konishi, T. Tamura, R. Zenitani, et al., Decline in energy storage in the Antarctic minke whale (Balaenoptera bonaerensis) in the Southern Ocean. Polar Biol. **31**, 1509–1520 (2008)

L. Krüger, Spatio-temporal trends of the Krill fisheries in the Western Antarctic Peninsula and Southern Scotia Arc. Fish. Manage. Ecol. **26**, 327–333 (2018)

L. Krüger, J.A. Ramos, J.C. Xavier, et al., Projected distributions of Southern Ocean albatrosses, petrels and fisheries as a consequence of climatic change. Ecography **41**, 195–208 (2018)

L. Krüger, M.F. Huerta, F. Santa Cruz, et al., Antarctic krill fishery effects over penguin populations under adverse climate conditions: implications for the management of fishing practices. Ambio **50**, 560–571 (2021)

J. Kuttippurath, P.J. Nair, The signs of Antarctic ozone hole recovery. Sci. Rep. **7**, 585 (2017)

A.F. Lacerda, L.S. Rodrigues, E. van Sebille, et al., Plastics in sea surface waters around the Antarctic Peninsula. Sci. Rep. **9**, 3977 (2019)

C. Lagger, M. Nime, L. Torre, et al., Climate change, glacier retreat and a new ice-free island offer new insights on Antarctic benthic responses. Ecography **41**, 579–591 (2018)

G. Lauriano, C.M. Fortuna, M. Vacchi, Occurrence of killer whales (Orcinus orca) and other cetaceans in Terra Nova Bay, Ross Sea, Antarctica. Antarct. Sci. 1 of 5 (2010)

R. Leaper, C. Miller, Management of Antarctic baleen whales amid past exploitation, current threats and complex marine ecosystems. Antarct. Sci. 1 of 27 (2011)

S.H. Lee, H.M. Joo, H.T. Joo, et al., Large contribution of small phytoplankton at Marian Cove, King George Island, Antarctica, based on long-term monitoring from 1996 to 2008. Polar Biol. **38**, 207–220 (2015)

J.R. Lee, B. Raymond, T. Bracegirdle, et al., Climate change drives expansion of Antarctic ice-free habitat. Nature **547**, 49–57 (2017)

E.W. Leuliette, R.S. Nerem, Contributions of Greenland and Antarctica to global and regional sea level change. Oceanography **29**, 154–159 (2016)

D. Liggett, B. Frame, N. Gilbert, et al., Is it all going south? Four future scenarios for Antarctica. Polar Rec. **53**, 459–478 (2017)

N. Liu, C.M. Brooks, China's changing position towards marine protected areas in the Southern Ocean: Implications for future Antarctic governance. Mar. Policy **94**, 189–195 (2018)

P. López-García, F. Rodríguez-Valera, C. Pedrós-Alió, Unexpected diversity of small eukaryotes in deep-sea Antarctic plankton. Nature **409**, 603–607 (2001)

F.L. Lowry, J.W. Testa, W. Calvert, Notes on winter feeding of Crabeater and Leopard Seals near the Antarctic Peninsula. Polar Biol. **8**, 475–478 (1988)

T. Maksym, Arctic and Antarctic sea ice change: contrasts, commonalities, and causes. Annu. Rev. Mar. Sci. **11**, 19.1–19.27 (2018)

M. Manganelli, F. Malfatti, T.J. Samo, et al., Major role of microbes in carbon fluxes during Austral Winter in the Southern Drake Passage. PLoS One **4**, e6941 (2009)

A. Martínez-Garcia, A. Rosell-Melé, E.L. McClymont, et al., Subpolar link to the emergence of the modern equatorial Pacific Cold Tongue. Science **328**, 1550–1553 (2010)

P.A. Mayewski, M.S. Twickler, S.I. Whitlow, et al., Climate change during the last deglaciation in Antarctica. Science **272**, 1636–1638 (1996)

A.H. McCarthy, L.S. Peck, Hughes, et al., Antarctica: the final frontier for marine biological invasions. Glob. Change Biol. **25**, 2221–2241 (2018)

J.B. McClintock, B.J. Baker, A review of the chemical ecology of Antarctic marine invertebrates. Am. Zool. **37**, 329–342 (1997)

J. McGee, M. Haward, Antarctic governance in a climate changed world. Austral. J. Maritime Ocean Aff. **11**, 78–93 (2019)

T. McIntyre, P.N.J. De Bruyn, I.J. Ansorje, et al., A lifetime at depth: vertical distribution of southern elephant seals in the water column. Polar Biol. **33**, 1037–1048 (2010)

J. Melbourne-Thomas, Climate shifts for krill predators. Nat. Clim. Change **10**, 386–391 (2020)

J. Melbourne-Thomas, S.P. Corney, R. Trebilco, et al., Under ice habitats for Antarctic krill larvae: could less mean more under climate warming? Geophys. Res. Lett. **43**, 10322–10327 (2016)

B. Meyer, L. Auerswald, V. Siegel, et al., Seasonal variation in body composition, metabolic activity, feeding, and growth of adult krill Euphausia superba in the Lazarev Sea. Mar. Ecol. Prog. Ser. **398**, 1–18 (2010)

T. Micol, P. Jouventin, Long term population trends in seven Antarctic seabirds. Polar Biol. **24**, 175–185 (2001)

K.G. Miller, M.A. Kominz, J.V. Browning, et al., The phanerozoic record of global sea-level change. Science **310**, 1293–1298 (2005)

K.A. Miller, K.F. Thompson, P. Johnston, et al., An overview of seabed mining including the current state of development, environmental impacts, and knowledge gaps. Front. Mar. Sci. **4**, 418 (2018)

M.A. Moline, H. Claustre, T.K. Frazer, et al., Alteration of the food web along the Antarctic Peninsula in response to a regional warming trend. Glob. Change Biol. **10**, 1973–1980 (2004)

S. Moreau, B. Mostjair, S. Bélanger, et al., Climate change enhances primary production in the western Antarctic Peninsula. Glob. Change Biol. **21**, 2191–2205 (2015)

C. Negre, R. Zahn, A.L. Thomas, et al., Climate change reverses Atlantic Ocean circulation. Nature **468**, 84–88 (2010)

S. Nicol, J. Foster, The fishery for Antarctic krill: its current status and management regime, in *Biology and Ecology of Antarctic Krill, Advances in Polar Ecology*, ed. by V. Siegel, (2016), pp. 387–421

S. Nicol, T. Pauly, N.L. Bindoff, et al., Ocean circulation off east Antarctica affects ecosystem structure and sea-ice extent. Nature **406**, 504–507 (2000)

T.D. Niederberger, E.M. Bottos, J.A. Sohm, et al., Rapid microbial dynamics in response to an induced wetting event in Antarctic dry valley soils. Front. Microbiol. **10**, 621 (2019)

C. Orejas, P. López-González, J.M. Gili, et al., Distribution and reproductive ecology of the Antarctic octocoral *Ainigmaptilon antarcticum* in the Weddell Sea. Mar. Ecol. Prog. Ser. **231**, 101–114 (2002)

C. Orejas, J.M. Gili, P. López-González, et al., Reproduction patterns of four Antarctic octocorals in the Weddell Sea: an inter-specific, shape, and latitudinal comparison. Mar. Biol. **150**, 551–563 (2007)

A. Pallesen, The legality of marine mining in the Antarctic treaty area, PhD Thesis, University of New Zealand, 2004

L.J. Pallin, S.C. Baker, D. Steel, et al., High pregnancy rates in humpback whales (*Megaptera novaeangliae*) around theWestern Antarctic Peninsula, evidence of a rapidly growing population. R. Soc. Open Sci. **5**, 180017 (2018)

D. Pardo, J. Forcada, A.G. Wood, et al., Additive effects of climate and fisheries drive ongoing declines in multiple albatross species. PNAS, E10829–E10837 (2017)

F. Pattyn, M. Morlighem, The uncertain future of the Antarctic Ice Sheet. Science **367**, 1331–1335 (2020)

T. Pavel, Russian researchers reach subglacial Lake Vostok in Antarctica. Adv. Polar Sci. **23**, 176–180 (2012)

D.A. Pearce, W.H. Wilson, Viruses in Antarctic ecosystems. Antarct. Sci. **15**(3), 319–331 (2003)

L. Peck, D.K.A. Barnes, A.J. Cook, et al., Negative feedback in the cold: ice retreat produces new carbon sinks in Antarctica. Glob. Change Biol. **16**, 2614–2623 (2010)

L.S. Peck, S.A. Morley, J. Richard, Acclimation and thermal tolerance in Antarctic marine ectotherms. J. Exp. Biol. **217**, 16–22 (2014)

S.E.A. Pineda-Metz, D. Gerdes, C. Richter, Benthic fauna declined on a whitening Antarctic continental shelf. Nat. Commun. **11**, 2226 (2020)

M.H. Pinkerton, J.M. Bradford-Grieve, Characterizing foodweb structure to identify potential ecosystem effects of fishing in the Ross Sea, Antarctica. ICES J. Marine Sci. **71**, 1542–1553 (2014)

A. Piñones, A.V. Fedorov, Projected changes of Antarctic krill habitat by the end of the 21st century. Geophys. Res. Lett. **43**, 8580–8589 (2016)

J. Plötz, H. Bornemann, R. Knust, et al., Foraging behaviour of Weddell seals, and its ecological implications. Polar Biol. **24**, 901–909 (2001)

J.O. Pope, P.R. Holland, A. Orr, et al., The impacts of El Niño on the observed sea ice budget of West Antarctica. Geophys. Res. Lett. **44**, 6200–6208 (2017)

H.O. Pörtner, C. Bock, F.C. Mark, Oxygen- and capacity-limited thermal tolerance: bridging ecology and physiology. J. Exp. Biol. **220**, 2685–2696 (2017)

R. Pudelko, P.J. Angiel, M. Potocki, et al., Fluctuation of glacial retreat rates in the Eastern part of Warszawa Icefield, King George Island, Antarctica, 1979–2018. Rem. Sens. **10**, 892 (2018)

M. Raes, A. Rose, A. Vanreusel, et al., Response of nematode communities after large-scale ice-shelf collapse events in the Antarctic Larsen área. Glob. Change Biol. **16**, 1618–1631 (2010)

R. Reese, G.H. Gudmundsson, A. Levermann, et al., The far reach of ice-shelf thinning in Antarctica. Nat. Clim. Change (2017). https://doi.org/10.1038/s41558-017-0020-x

C.S. Reiss, A. Cossio, J.A. Santora, et al., Overwinter habitat selection by Antarctic krill under varying sea-ice conditions: implications for top predators and fishery management. Mar. Ecol. Prog. Ser. **568**, 1–16 (2017)

E. Rignot, H. Thomas, Mass balance of polar ice sheets. Science **297**, 1502–1506 (2002)

E. Rignot, J. Mouginot, B. Scheuchl, et al., Four decades of Antarctic ice sheet mass balance from 1979–2017. PNAS **116**, 1095–1103 (2019)

S.R. Rintoul, S.L. Chown, R.M. DeConto, et al., Choosing the future of Antarctica. Nature **558**, 233–241 (2018)

A.D. Rogers, B.A.V. Frinault, D.K.A. Barnes, et al., Antarctic futures: an assessment of climate-driven changes in ecosystem structure, function, and service provisioning in the Southern Ocean. Annu. Rev. Mar. Sci. **12**, 7.1–7.34 (2020)

J. Roman, J.J. McCarthy, The whale pump: marine mammals enhance primary productivity in a coastal basin. PLoS One **5**, e13255 (2010)

Y. Ropert-Coudert, A. Kato, X. Meyer, et al., A complete breeding failure in an Adélie penguin colony correlates with unusual and extreme environmental events. Ecography **37**, 001–003 (2014)

A. Rose, J. Ingels, M. Raes, et al., Long-term iceshelf-covered meiobenthic communities of the Antarctic continental shelf resemble those of the deep sea. Mar. Biodivers. **45**, 743–762 (2015)

S. Rossi, *Oceans in Decline* (Copernicus Books, Springer-Nature, 2019), 349 pp

S. Rossi, F. Elias-Piera, Trophic ecology of three echinoderms in the deep waters of the Weddell Sea (Antarctica). Mar. Ecol. Prog. Ser. **596**, 143–153 (2018)

S. Rossi, L. Rizzo, Marine animal forests as C immobilizers or why we should preserve these three-dimensional alive structures, in *Perspectives on the Marine Animal Forests of the World*, ed. by S. Rossi, L. Bramanti, (Springer-Nature, 2020), pp. 333–399

S. Rossi, E. Isla, A. Martínez-García, et al., Transfer of seston lipids during a flagellate bloom from the surface to the benthic community in the Weddell Sea. Sci. Marina **77**, 397–407 (2013)

S. Rossi, L. Bramanti, A. Gori, et al., An overview of the animal forests of the world, in *Marine Animal Forests: The Ecology of Benthic Biodiversity Hotspots*, ed. by S. Rossi, L. Bramanti, A. Gori, C. Orejas, (Springer, 2017a), pp. 1–28

S. Rossi, M. Coppari, N. Viladrich, Benthic-Pelagic coupling: new perspectives in the animal forests. In: Marine Animal Forests: The Ecology of Benthic Biodiversity Hotspots, in , ed. by S. Rossi, L. Bramanti, A. Gori, C. Orejas, (Springer, Berlin, 2017b), pp. 855–886

S. Rossi, E. Isla, M. Bosch-Belmar, et al., Changes of energy fluxes in the marine animal forest of the Anthropocene: factors shaping the future seascape. ICES J. Mar. Sci. **76**, 2008–2019 (2019)

H. Rott, P. Skvarca, T. Nagler, Rapid collapse of Northern Larsen Ice Shelf, Antarctica. Science **271**, 788–792 (1999)

L.G. Sancho, A. Pintado, T.G.A. Green, Antarctic studies show lichens to be excellent biomonitors of climate change. Diversity **11**, 42 (2019)

T. Sandersfeld, F.C. Mark, R. Knust, Temperature-dependent metabolism in Antarctic fish: do habitat temperature conditions affect thermal tolerance ranges? Polar Biol. **40**, 141–149 (2017)

E. Sañé, E. Isla, A. Grémare, et al., Pigments in sediments beneath recently collapsed ice shelves: the case of Larsen A and B shelves, Antarctic Peninsula. J. Sea Res. **65**, 94–102 (2011)

T.A. Scambos, R.E. Bell, R.B. Alley, et al., How much, how fast?: a science review and outlook for research on the instability of Antarctica's Thwaites Glacier in the 21st century. Glob. Planet. Sci. **153**, 16–34 (2017)

A. Schiavone, K. Kannan, Y. Horii, et al., Occurrence of brominated flame retardants, polycyclic musks, and chlorinated naphthalenes in seal blubber from Antarctica: Comparison to organochlorines. Marine Pollut. Bull. **58**, 1406–1419 (2009)

W.H. Schlesinger, Global change ecology. Trends Ecol. Evol. **21**, 348–351 (2006)

O. Schofield, M. Brown, J. Kohut, et al., Changes in the upper ocean mixed layer and phytoplankton productivity along the West Antarctic Peninsula. Philos. Trans. R. Soc. A **376**, 20170173 (2018)

N. Servetto, S. Rossi, V. Fuentes, et al., Seasonal trophic ecology of the dominant Antarctic coral Malacobelemnon daytoni (Octocorallia, Pennatulacea, Kophobelemnnidae). Mar. Environ. Res. **130**, 264–274 (2017)

A.A. Sfriso, Y. Tomio, B. Rosso, et al., Microplastic accumulation in benthic invertebrates in Terra Nova Bay (Ross Sea, Antarctica). Environ. Int. **137**, 105587 (2020)

V. Siegel, Distribution and population dynamics of Euphausia superba : summary of recent findings. Polar Biol. **29**, 1–22 (2005)

M. Siegert, A. Atkinson, A. Banwell, et al., The Antarctic Peninsula under a1.5°C global warming scenario. Front. Environ. Sci. **7**, 102 (2019)

G.K. Silber, M.D. Lettrich, P.O. Thomas, et al., Projecting marine mammal distribution in a changing climate. Front. Mar. Sci. **4**, 413 (2017)

J. Singh, R. Singh, R. Khare, Influence of climate change on Antarctic flora. Polar Sci. **18**, 94–101 (2018)

V. Smetacek, S.W.A. Naqvi, The next generation of iron fertilization experiments in the Southern Ocean. Philos. Trans. R. Soc. A **366**, 3947–3967 (2008)

V. Smetacek, S. Nicol, Polar ocean ecosystems in a changing world. Nature **437**, 362–368 (2005)

V. Smetacek, P. Assmy, J. Henjes, The role of grazing in structuring Southern Ocean pelagic ecosystems and biogeochemical cycles. Front. Mar. Sci. **16**, 541–558 (2004)

V. Smetacek, C. Klaas, V.H. Strass, et al., Deep carbon export from a Southern Ocean iron-fertilized diatom bloom. Nature **487**, 313–319 (2012)

K.L. Smith Jr., B.H. Robison, J.J. Helly, et al., Free-drifting icebergs: hot spots of chemical and biological enrichment in the Weddell Sea. Science **317**, 478–482 (2007)

S. Solomon, R. Stearns, On the role of the weather in the deaths of R. F. Scott and his companions. PNAS **9**, 13012–13016 (1999)

S. Solomon, R.W. Portmann, D.W.J. Thompson, Contrasts between Antarctic and Arctic ozone depletion. PNAS **104**, 445–449 (2007)

P.T. Sontag, D.K. Steinberg, J.R. Reinfelder, Patterns of total mercury and methylmercury bioaccumulation in Antarctic krill (Euphausia superba) along the West Antarctic Peninsula. Sci. Total Environ. **688**, 174–183 (2019)

S.E. Stammerjhon, T. Maksym, R.A. Masson, et al., Seasonal sea ice changes in the Amundsen Sea, Antarctica, over the period of 1979–2014. Elementa **3**, 000055 (2015)

E.J. Steig, P.D. Neff, The prescience of paleoclimatology and the future of the Antarctic ice sheet. Nat. Commun. **9**, 2730 (2018)

E.J. Steig, E.J. Brook, J.W.C. White, Synchronous climate changes in Antarctica and the North Atlantic. Science **282**, 92–95 (1998)

B.B. Stephens, R.F. Keeling, The infuence of Antarctic sea ice on Glacial interglacial CO_2 variations. Nature **404**, 171–174 (2000)

E. Stokstad, Boom and bust in a polar hot zone. Science **315**, 1522–1523 (2007)

C.W. Sullivan, K.R. Arrigo, C.R. McClain, et al., Distributions of phytoplankton blooms in the Southern Ocean. Science **262**, 1832–1837 (1993)

M. Sun, Antarctica pact could open way for mining. Science **240**, 1612 (1988)

W.J. Sutherland, R. Aveling, T.M. Brooks, et al., A horizon scan of global conservation issues for 2010. Trends Ecol. Evol. **25**, 1–7 (2010)

S. Taboada, L.F. García-Fernández, S. Bueno, et al., Antitumoural activity in Antarctic and sub-Antarctic benthic organisms. Antarct. Sci. **22**(5), 494–507 (2010)

S. Taniguchi, R.C. Montone, M.C. Bicego, et al., Chlorinated pesticides, polychlorinated biphenyls and polycyclic aromatic hydrocarbons in the fat tissue of seabirds from King George Island, Antarctica. Mar. Pollut. Bull. **58**, 129–133 (2009)

N. Teixidó, J. Garrabou, J. Gutt, et al., Recovery in Antarctic benthos after iceberg disturbance: trends in benthic composition, abundance and growth forms. Mar. Ecol. Prog. Ser. **278**, 1–16 (2004)

S. Thatje, The future fate of the Antarctic marine biota? Trends Ecol. Evol. **8**, 418–419 (2005)

S. Thatje, C.D. Hillenbrand, A. Mackensen, et al., Life hung by a thread: endurance of Antarctic fauna in glacial periods. Ecology **89**, 682–692 (2008)

S. Thatje, A. Brown, C.D. Hillebrand, Prospects for metazoan life in sub-glacial Antarctic lakes: the most extreme life on Earth? Int. J. Astrobiol. **18**, 416–419 (2019)

D. Thomas, *Frozen oceans* (Natural History Museum, London, 2004), 224 pp

E. Thomas, An ocean view of the early cenozoic: greenhouse world. Oceanography **19**, 94–103 (2006)

D. Thomas, G.E. Fogg, P. Convey, et al., *The biology of polar regions* (Oxford University Press, 2008), 394 pp

P.N. Thrathan, P. Garcia-Borboroglu, D. Boersma, et al., Pollution, habitat loss, fishing, and climate change as critical threats to penguins. Conserv. Biol. **29**, 31–41 (2014)

S.F. Thrush, J.E. Hweitt, V.J. Cummings, et al., β-diversity and species accumulation in Antarctic Coastal Benthos: influence of habitat, distance and productivity on ecological connectivity. PLoS One **5**, e11899 (2010)

T. Tin, Z.L. Fleming, K.A. Hughes, et al., Impacts of local human activities on the Antarctic environment. Antarct. Sci. **21**(1), 3–33 (2009)

R. Traczyk, V.B. Meyer-Rochow, R.M. Hughes, Icefish adaptations to climate change on the South Georgia Island Shelf (Sub-Antarctic). Ocean Sci. J. **55**, 303–319 (2020)

P. Tréguer, G. Jacques, Dynamics of nutrients and phytoplankton, and fluxes of carbon, nitrogen and silicon in the Antarctic Ocean. Polar Biol. **12**, 149–162 (1992)

V.J.D. Tulloch, E.E. Pláganyi, C. Brown, et al., Future recovery of baleen whales is imperiled by climate change. Global Change Biol. **25**, 1263–1281 (2016a)

V.J.D. Tulloch, E.E. Pláganyi, R. Matear, et al., Ecosystem modelling to quantify the impact of historical whaling on Southern Hemisphere baleen whales. Fish Fish. **19**, 117–137 (2016b)

M. Van Caspel, M. Schröder, O. Huhn, et al., Precursors of Antarctic bottom water formed on the continental shelf off Larsen Ice Shelf. Deep-Sea Res. I **99**, 1–9 (2015)

D.G. Vaughan, R. Arthern, Why is it hard to predict the future of ice sheets? Science **315**, 1503–1504 (2007)

W.F. Vicent, Evolutionary origins of Antarctic microbiota: invasion, selection and endemism. Antarct. Sci. **12**(3), 374–385 (2000)

J.E. Walsh, A comparison of Arctic and Antarctic climate change, present and future. Antarct. Sci. **21**(3), 179–188 (2009)

H.S. Wauchope, J.D. Shaw, A. Terauds, A snapshot of biodiversity protection in Antarctica. Nat. Commun. **10**, 946 (2019)

B. Wienecke, Review of historical population information of emperor penguins. Polar Biol. (2010)

R. Williams, N. Kelly, O. Boebel, et al., Counting whales in a challenging, changing environment. Sci. Rep. **4**, 4170 (2014)

G. Williams, T. Maksym, J. Wilkinson, et al., Thick and deformed Antarctic sea ice mapped with autonomous underwater vehicles. Nat. Geosci. **8**, 61–67 (2015)

J.K. Willis, D.P. Chambers, C.Y. Kuo, et al., Global Sea level rise. Oceanography **23**, 26–35 (2010)

J.C. Xavier, A. Barbosa, S. Agusti, et al., Polar marine biology science in Portugal and Spain: recent advances and future perspectives. J. Sea Res. **83**, 9–29 (2013)

X. Xiong, L. Guardone, J.M. Cornax, et al., DNA barcoding reveals substitution of Sablefish (Anoplopoma fimbria) with Patagonian and Antarctic Toothfish (Dissostichus eleginoides and Dissostichus mawsoni) in online market in China: how mislabeling opens door to IUU fishing. Food Contr. **70**, 380–391 (2016)

K. Xu, F.X. Fu, Hutchins, Comparative responses of two dominant Antarctic phyto-plankton taxa to interactions between ocean acidification, warming, irradiance, and iron availability. Limnol. Oceanogr. **59**, 1919–1931 (2014)

X. Yuan, ENSO-related impacts on Antarctic sea ice: a synthesis of phenomenon and mechanisms. Antarct. Sci. **16**, 415–425 (2004)

N. Yuan, M. Ding, Y. Huang, et al., On the long-term climate memory in the surface air temperature records over Antarctica: a nonnegligible factor for trend evalua-tion. J. Clim. **28**, 5922–5934 (2015)

Printed in the United States
by Baker & Taylor Publisher Services